T0353495

A Passion for
Discovery

A Passion for
Discovery

Peter Freund

World Scientific

NEW JERSEY · LONDON · SINGAPORE · BEIJING · SHANGHAI · HONG KONG · TAIPEI · CHENNAI

Published by

World Scientific Publishing Co. Pte. Ltd.

5 Toh Tuck Link, Singapore 596224

USA office: 27 Warren Street, Suite 401-402, Hackensack, NJ 07601

UK office: 57 Shelton Street, Covent Garden, London WC2H 9HE

British Library Cataloguing-in-Publication Data
A catalogue record for this book is available from the British Library.

A PASSION FOR DISCOVERY

ISBN-13 978-981-270-646-1
ISBN-10 981-270-646-1
ISBN-13 978-981-277-214-5 (pbk)
ISBN-10 981-277-214-6 (pbk)

Typeset by Stallion Press
Email: enquiries@stallionpress.com

Printed in Singapore.

Preface

Beyond its intrinsic beauty, the pursuit of physics, now into its fifth century, comes with a thrilling and compelling human story. From its earliest beginnings — think Galileo — this story has fascinated physicists and non-physicists alike. Over the years, I both heard and got to witness many such stories. They put a human face on the grand story, the evolution of our science. As a rule these stories are handed down by word of mouth. After finishing the proof of Emmy Noether's seminal theorem in my field theory course, I would let the students catch their breath, while telling them the moving story of this great mathematician. Invariably they were riveted; you could hear a pin drop. The idea of putting such stories down on paper for a change, was born, with the hope of conveying the excitement and flavor of the physicist's life, as well as the way physicists' lives are affected by the sometimes benign, at other times malignant, forces of history.

Though what follows is built out of many, many anecdotes, vignettes, and stories, they naturally cluster around certain clearly visible main themes, which give a degree of continuity to the whole. *What* all these scientists try to achieve, their deeper motivation, is obviously of essence, and, rigorously foreswearing the use of any technical means, I have tried to give at least a general idea of the beautiful goal of the physicist's work. Above all, this book is about what happens at the human level when some extremely gifted individuals, living in the unavoidable historic reality, devote themselves to what can be called without hyperbole a transcendental goal.

By now I have twice referred to the beauty of physics and I ought to explain my use of this word. One normally associates beauty with works of art. It takes little to explain why Jean-Auguste-Dominique Ingres' *Grande Odalisque,* on view at the Louvre, is beautiful. The shapes are superb, the color scheme rich and sensual; the varnished texture intimates perfection. It all seems so inevitable, so elegant. It is these two ingredients, inevitability and elegance, or economy of means, which we find as well in a major scientific idea. In physics, the ultimate focus is on how many ideas previously thought of as unrelated now get connected into a meaningful whole and become much easier to comprehend, and then what entirely new perspectives open up before us.

Before Einstein's theory of general relativity, for example, the shape of space (or more generally, the shape of the four-dimensional space–time continuum) had to be postulated, or, more frankly, guessed. Then, in this guessed space one could study the problem of how gravity is generated by matter, and in turn affects the movement of matter. The shape of space had neither much to do with this problem, nor was it affected by this problem. But when the theory of general relativity was formulated and then experimentally confirmed, the interaction between matter and gravity was seen to also *determine* the shape of space, and guesswork was no longer needed. Two independent problems turned out to reflect different aspects of a larger whole. As a consequence, the evolution of space could be studied and modern cosmology was born. How bold, how inevitable, how economic! How beautiful!

I have used a few criteria in selecting the stories in this book. I tried to exclude as much as possible anything having to do with my own work, lest some ulterior motives be read into my writing. I also steered away from stories that are widely known, such as Feynman's participation in the investigation of the Challenger disaster and Heisenberg's wartime meeting with Bohr, etc. I have tried to conscientiously check the stories I am telling. To a large extent I relied on my own good, though not infallible, memory. If there are some inaccuracies in this book, I am sorry and stand ready to be corrected. I hope, however, that my love and deep respect for the people I write about shines through these pages. There are some "heavies" in the book, at least as far as my narrative goes. But even these would not be worth writing about if they weren't as remarkable, great, and influential as they unquestionably are.

Quite often I try to understand the mindset of one or another of the people I write about. Some of my interpretations, I am sure, will give rise to controversy, but that is as it should be.

All verbatim quotations in these pages are my renderings of what I remember as having been said, or what I remember having been told was said. I sincerely hope that I have gotten the substance of these quotations right and that the words I place in peoples' mouths fit comfortably and deliciously, even if on occasion a bit painfully, there.

This book is about physics and about narratives. At first one might think the two have nothing in common. Yet, there is a close connection. To a large extent, to understand something is to be able to put it in narrative form. It should then come as no surprise if we find narrative structures in physics. In fact there are three narrative structures running through physics.

First, each physics or mathematics paper has its own narrative structure: the proofs are not instantaneous, they develop in time. One can get as glued to a physics or mathematics paper as to a good mystery story. The heroes are the various *concepts*. Their adventures wring them through all those equations and at the end they emerge changed, or they even beget new concepts. Of course, such a narrative also involves minor characters, and we are often left wondering what happened to them. At the end of the paper, not surprisingly, we often find a throwaway remark like, "We hope to return to this elsewhere," and sometimes the paper is followed up by what in Hollywood would be called a sequel, or a spin-off.

On the second level, there is a narrative running through all of physics: the magnificent story of this science in its entirety. In this narrative the various *theories* are the heroes, as they change and ultimately evolve into better or altogether new ones. Each time an important new paper is written, all old papers implicitly get rewritten and we can explain to the young in a matter of years what took centuries to discover. Similar things happen in the arts. Before starting to write, a novelist does not have to read all the novels ever written, starting with Longus and Petronius, for at some level each new novel contains and also reacts to the wisdom of its ancestors. Once Tolstoy wrote *Anna Karenina*, Emma Bovary acquired company she would never be able to shake.

Finally, there is a third level at which physics has a narrative structure: the human level. Like all science or art, physics is a human enterprise, and its practitioners strongly interact with each other. Under the right circumstances these interactions result in brilliant work and beautiful friendships. But there is more to it. Physicists do not live in an ivory tower; they are not spared the ravages of history. Very dramatic situations can be whipped up by the, often irrational, winds of history and politics. This drama, such as I have had the chance to observe it, is the focus of the stories I tell here.

In my life I have had ample occasion, more indeed than I could have wished for, to witness drama served up by history, which leaves its indelible stamp on all humans and in particular on scientists, be they physicists, mathematicians, biologists.... In Romania, where I grew up and got my undergraduate education, after the pleasures of a wartime fascist dictatorship, we were presented the glories of the dictatorship of the proletariat, in its purest Stalinist guise. The 1956 Hungarian uprising led to events in my hometown of Timişoara, during which I, an undergraduate at the Polytechnic Institute there, found myself lined up with other students between a wall and a line of tanks with their guns pointed at us. Obviously, for some rather bizarre — if for us happy — reasons, the communist authorities decided against mowing us down. But as far as I am concerned, my sensibility to political upheaval has forever been considerably heightened. I developed a keen interest in the way the physicists I came in contact with, managed to cope with the kind of crises that the twentieth century so copiously lavished on the supposedly civilized world.

During my graduate studies at the University of Vienna, my post-doctoral appointments at the University of Geneva and the Institute for Advanced Study in Princeton, and my many years as a professor of Physics at the University of Chicago, I got a chance to participate in and contribute to the development of the ideas that culminated in today's all-encompassing string theory. All through these years, I was in close contact with physicists whose own lives bore the distinct impact of the twentieth century's grand events, and who in turn had known others for whom this was true. A picture, a narrative, started forming in my mind, which organized these remarkable human stories in a meaningful whole. Interestingly, this overarching human story is inextricably intertwined with the narrative of physics itself, lending a contrapuntal structure to this book.

Peter Freund
Chicago
Spring, 2007

Acknowledgments

The idea of writing something along these lines was first suggested to me by the English mathematical physicist Nick Manton quite some time back. At that point I had in mind writing for younger theoretical physicists to make sure these stories do not get lost over the years. More recently, my daughter Caroline Freund and my son-in-law Simeon Djankov suggested something similar but with a different typical reader in mind. They made it clear to me that these stories are not without interest for non-physicists. Once my fingers started banging the keyboard, many friends and in some cases almost-former friends gave of their time to read parts of the emerging typescript, giving me what I believe was their frank reaction.

My wife Lucy had to put up with my reading to her each chapter fresh out of the printer. This was extremely useful, since hearing her laugh and gasp convinced me that whatever sins I may be guilty of, the mortal sin of being boring was not among them.

It was a privilege and a great pleasure to have my teacher Walter Thirring read the typescript and comment on it. To him I owe so much.

My special thanks go to David Fairlie, Wally Greenberg, Emil and Joann Martinec, Yoichiro Nambu, Reinhard Oehme and the late Mafalda Oehme, Jacqui Sanders and that unique and much-missed physicist, the late Valentine Telegdi, who gave of their time and made extensive and very useful comments, which helped me improve, I hope, the accuracy and intelligibility of the text. In some instances they also contributed material, which they kindly permitted me to include.

My good friend John Ryden's assessments of what in the first draft worked and what did not, was more useful than I could possibly acknowledge here. Without his input this book would have missed the goal it strives to reach.

It is a pleasure to extend my gratitude and thanks to many friends, in particular to Laurie and Brigitte Brown, Thomas and Jo Ann Curtright, Henry Frisch, Charles J. Goebel, Luca Mezincescu, Rafael and Susana Nepomechie, Mircea Pigli, Jon and Joy Rosner, John Szöke, Ben and Anita Wolfe, and Ira Wool, for their useful comments and for their support.

My sincere thanks go out to Sultan Catto and Brian Schwartz for putting me in touch with Dorian Devins of World Scientific, with whom I discovered a vast communality of intellectual interests and to whom I am deeply indebted.

To Ian Seldrup, of World Scientific, I owe deep thanks for his very valuable editorial help.

I also gratefully acknowledge Ms. Janet Young's early editorial assistance.

I wish to thank, without naming them, the many physicists and mathematicians I interacted with over the past four decades, be it as collaborators, as teachers, as students, as kindred spirits, as role models, as competitors. It is due to these intense interactions that I was in the proper frame of mind to write this book.

Last but not least, to my marvelous immediate family, to Lucy Freund, Pauline Freund, Caroline Freund, John Smither, Simeon Djankov, Maddie Smither, Raphi Smither, Andrei Freund, Charlotte Smither, and Marko Freund, I am deeply indebted for their much-needed moral and emotional support during the writing of this book.

Contents

Einstein Once Removed

It was late fall when my colleague Subramanyan Chandrasekhar — Chandra as we all called him — usually the epitome of aristocratic demeanor, appeared in my office, visibly shaken up, indeed outraged.

"Have you seen the new Schrödinger biography?" He was referring to Walter Moore's delightful book about the great Austrian physicist Erwin Schrödinger. "It contains pictures of five of his mistresses and two of his illegitimate daughters."

"So what? Physicists are not saints, and for that matter, even saints are not always saints. Saint Augustine?"

"Oh, but Peter, when I was a student, the great teacher Sommerfeld came to India to talk about the new quantum mechanics. He impressed upon me the fact that this was the creation of a group of exceptional young men: Heisenberg, Schrödinger, Dirac and Pauli, who had dedicated themselves to science to the exclusion of everything else in life. There and then I vowed to follow their example and I have lived accordingly all my life, and now this...."

I was so intrigued by what Chandra had said that I immediately headed for the bookstore and purchased the book. That Schrödinger had been a ladies' man came as no big surprise to me; stories about his dalliances are legion. What came as a real shock for me was to learn that his wife, Annemarie, had had an affair with Hermann Weyl, one of the twentieth century's greatest mathematicians — Alma Mahler-Gropius-Werfel eat your heart out. I had met Mrs. Schrödinger during my student days in Vienna. On more than one occasion after her illustrious husband's death, I had lunch at her IX-th district apartment. She was a native of Salzburg where she grew up next door to those "insufferable Karajans." Mrs. Schrödinger did not take well to widowhood. She got quite depressed. At the recommendation of friends and of her doctor she frequently invited guests for lunch, in Austria still the main meal of the day. As it happened, she had the services of one of Vienna's finest cooks. In all of Vienna you could not eat better *Beuschel mit Knödel*, that quintessentially Austrian delicacy made of calf's lungs. It is at one of these lunches that I first heard about the Einstein–Schrödinger feud, which grew out of their unified theory.

Unified theories have grown out of the realization that many of the forces and forms of matter that play an important role in our lives (such as friction, pushing, pulling, doors, fire, water, etc.) are but irrelevancies in the grand scheme of things. It all boils down to a very small number of *fundamental* forces and *fundamental* forms of matter. Just as the ancients kept reducing the number of fundamental deities, till in the end they landed at the doorstep of one or of another form of monotheism, so in this striving to understand nature, a reductive struggle also goes on and ultimately brings us to the doorstep of one or of another unified theory.

When introducing words like *unified theory*, I should admit that over the centuries, physicists have been rather profligate in the use of language. Words like *atom* and *proton* have been coined in the excitement of the moment, only to be superseded in the quest for the *truly* fundamental forms of matter. The same with forces, for which unified theories and then even *grand* unified theories have been constructed only to yield later on to the more fundamental as well. But then, all these grandiose words bear witness to the heroism of the quest and the dangers of letting oneself get carried away by what parades before us in the guise of triumph. It is a little like walking around some European capital and suddenly being hit by the realization that many of those generals and marshals for whom the streets are named ultimately lost their wars.

The very first step on the road to unification was taken in the nineteenth century when James Clerk Maxwell unified electricity with magnetism in a beautiful and coherent theory. That Maxwell's example should be heeded and that *all* forces should be unified was first proposed by Gustav Mie early in the twentieth century, though at that time all forces hadn't even been discovered yet. Mie was the Giacomo Meyerbeer of unified theory. Just as that composer's operas are now almost forgotten and were never taken seriously by Verdi and Wagner, the heavyweights of grand opera in Meyerbeer's day, so Mie's work on unification is now essentially forgotten and was never taken seriously by Einstein. Things even went so far that Wagner, motivated in part by his anti-Semitism, openly proclaimed his outright hatred for Meyerbeer, and that Einstein's gut reaction to Mie, though untainted by any bigotry, was also heavily antagonistic. Yet, just as *grand* opera — do I detect verbal profligacy among the musicians as well? — would be inconceivable without Meyerbeer, so it is doubtful that the urge to unify would have struck as early as it did without Mie.

In the middle of the twentieth century Albert Einstein in Princeton and Erwin Schrödinger in Dublin were independently at work on an early variety of a unified theory, in which only gravity and electromagnetism are considered worth unifying. The two men were utterly unfamiliar with the remaining two fundamental forces — the, at that time very poorly understood, weak and strong forces responsible for beta-radioactivity and nuclear energy. These limitations were further compounded by the fact that the sixtyish Schrödinger and the seventyish Einstein came up with a macroscopic, or "classical" theory, ill-defined at the microscopic level of atoms and of their constituents. At that level energy is emitted in quanta and a new different "quantum theory" applies.

The two physicists constructed this theory in a leisurely fashion and exchanged letters on its details. Then, in 1947 Schrödinger gave a lecture about it, and made some exaggerated claims as to its prospects. These were reported in the press much to Einstein's dismay. In a *folie à deux* the two old physicists greatly overestimated the value of this incomplete unification. Older wartime issues, over which the two former Berlin colleagues had had an earlier falling out, fanned the flames. This much is well known and excellently told in Moore's book. But there is more to the story. According to Mrs. Schrödinger, the Einstein–Schrödinger feud escalated to the point where the ugly p-word — plagiarism — was sounded by both parties, and both were considering legal action. It is at this moment that the right mediator, Wolfgang Pauli, stepped in. Pauli was respected by both unifiers as brilliant, critical, thoroughly honest and frank. No matter what great things they had done in the past, he warned them, were they to sue each other, they would become the laughing stock of the whole world and their reputations would badly suffer. He then added, "Besides, I really don't see what the whole fuss is about. This theory is ill conceived. If you connected *my* name with it in any fashion then *I* would have a right to sue *you*." This did the trick.

Old age goes a long way towards explaining this episode; older people tend to be more suspicious and less critical. Add to this that in old age Werner Heisenberg, Einstein's only true peer at the very pinnacle of twentieth-century physics, also went about finding *his,* you guessed it, unified theory, initially in collaboration with none other than Wolfgang Pauli, who to his credit pulled out before anything got published.

What is it that made all these giants chase after unification in old age? One can answer this question on two levels. On one level it is an act of ultimate grandiosity. All these men were time and again extremely successful in physics. As they see their powers waning, they take one final stab at the biggest problem: finding the ultimate theory, *ending* physics. To do so is somewhat egotistic, for if successful, all that would be left to later generations would be to mop up and move on — to the study of complex systems, of chaos, wherever, but away from the search for beautiful simple laws, as these would all have been found. Maybe aside from their property of being readily mapped onto the human brain, the laws of nature are such that it is impossible to find them all, or at least impossible to know that one has found them all. This way, full employment for future generations of physicists would be guaranteed.

On another level, maybe these men are just driven by the same insatiable curiosity that has stood them in such good stead in their youth. They want to know the solution to the puzzle that has preoccupied them throughout life; they want to have a glimpse of the promised land in their lifetime. I don't pretend to know their true motives: the fact is, what all these masters (along with Einstein, Schrödinger and Heisenberg, I include Mie, Nordström, Kaluza, Klein and Weyl) have started, has become the main preoccupation of a whole new generation, still hoping for that TOE — that Theory Of Everything. Maybe it is a mirage, but maybe it is truly within our grasp.

Heisenberg's Turn

Barely a decade after the Einstein–Schrödinger debacle, Heisenberg embarked on his version of a unified theory. Still in his fifties, his was a different, somewhat more youthful creation, intended as a full-fledged microscopic quantum theory but ignoring gravity, as if the apple hadn't fallen on Newton's head. Like his predecessors, Heisenberg attempted to unify not everything but *almost* everything. Where they differed is in what they left out. Einstein–Schrödinger made sure their apples were falling as they should, even as they left radioactivity and nuclear energy out of their macroscopic theory.

Heisenberg's attempt was flawed, it led to the preposterous prediction that the probability of certain physical processes occurring would have to be negative. If you think about that for a moment, this is a fatal flaw. If you are sure that something *will* occur, its probability is one (100 percent); if you're sure it *won't* occur, its probability is zero. A probability cannot exceed one, you cannot be surer than sure. In the same way, a probability cannot be smaller than zero, for then you'd be surer than sure that it won't occur. But all negative numbers are by definition smaller than zero, so a theory that predicts negative probabilities is nonsense. Heisenberg understood this difficulty, but hoped to somehow get around it. He never did, but not for lack of trying. At this point we could simply dismiss his proposal. Yet unlike Einstein and Schrödinger's earlier work, Heisenberg's unification, flawed though it was, raised some issues that led to a major development. Specifically, he asked himself how symmetries are broken in nature. He then made a truly inspired analogy with ferromagnetism. In our understanding of ferromagnetism, the magnetism of an iron magnet, the laws have no preferred direction built in; they treat all directions alike in a democratic fashion. This is an example of symmetry, rotational symmetry at that. By going around we see no change in these laws: they are symmetric, like a perfectly polished sphere that looks the same from all directions. Yet by the time we are done, we end up with a magnetic field which of necessity has to point in some direction. This violates the democracy of the laws from which we started, as if someone had dented the sphere so that the direction from its center to the dent can be distinguished from all other directions. How did this preferred direction arise? How was the democracy violated, or as we put it, how was the symmetry broken? More about that later.

For the time being let us ask how it is that Heisenberg, after raising the important question of how symmetries are broken in nature and making this analogy, managed to miss all the truly remarkable physics it led to. He was obsessed, or so it seems, with the many irrelevant details of his theory. The eagle's vision of his earlier years had gotten badly blurred. Age? He was in his fifties. Though considerably younger than Einstein and Schrödinger when they worked on unification, age could be part of the story, but there had to be more to it. The post-World-War-II Heisenberg was a different man from the prewar Heisenberg. His creative powers had taken a dive during the war, when he was still in his early forties. This brings us to the murky issue of Heisenberg's position towards the Third Reich. A lot of ink has been spilled over this issue — essays, memoirs, even a fine play. Most of these writings deal with the "what did Heisenberg know and when did he know it?" type of questions. At this late time there is little anyone can add at that level. To my mind there are other more relevant, interesting and important questions. *Why* did Heisenberg do whatever he did? In his own mind did he end up viewing what he had done as ethically proper or ethically flawed? Finally, how did all this stop his creative juices from flowing? On these questions there is evidence which bears further consideration.

To begin with, let me forcefully state that Heisenberg was one of a handful of the greatest physicists of all time, a Picasso of twentieth century physics. But where does that leave Einstein? Well, unlike the arts, physics produced *two* Picassos in the twentieth century. It can happen. Think of music between the middles of the eighteenth and of the nineteenth centuries: Mozart *and* Beethoven *and* Schubert. Heisenberg gave us quantum mechanics, the framework for describing the microscopic world. He was the first to understand the deep epistemological changes this theory called for, and it was he who both produced the most spectacular applications of the new theory and opened the most significant paths along which the theory should be developed. It is fair to say that worldwide from 1925 to 1943 he was the dominant figure in theoretical physics.

Heisenberg did his last great work on the mathematical and physical riches stored in the so-called S-matrix, at the height of the war in 1943. With his characteristic conceptual clarity, he asked for the right mathematical object which encodes everything we could conceivably measure in the realm of validity of a given microscopic quantum theory. He then defined this object, the S-matrix, and proved some of its truly remarkable properties. Justifiably, he got so excited about this work that he arranged for a copy of his paper to be delivered to Japan by German submarine. In 1943 he was only forty-two years old. But of course 1943 was but two years before Hitler's defeat. Had Heisenberg, like so many others, come to the States, he would probably have continued working at peak form for quite some years after the war. His friend Enrico Fermi and many others urged him to leave Germany, but he decided to stay. He saw the war coming, but felt a duty towards the young German physicists and this compelled him to stay. But to stay meant finding a way to make his peace with Hitler. He did.

Heisenberg was an erudite man with an intense philosophic bent, and he did not reach his decision lightly; he searched his soul for months before making up his mind on whether he should stay in Germany during the war. He has written extensively about this struggle. As I understand it, he reasoned that Stalin's Russia presented a mortal danger to German culture and had to be dealt with. If it took a Hitler to do the job, so be it. In other words, he did not see Hitler as evil incarnate, but as a historic figure whom he despised and whose fanaticism disgusted him, but who nevertheless had a role to play. After defeating Stalin, Hitler would be disposed of one way or another — Heisenberg was sure of that — and things in Germany would return to business as usual. He felt he had no alternative but to support his fatherland in its hour of need. Indeed, it was a *moral imperative*.

After the war, many of his close friends outside Germany, people he respected, took him to task for this decision, which *they* found morally flawed. He was bewildered, having based his decision on clear, rigorous reasoning. But no-one was faulting his reasoning; people faulted his assumptions and his result. He went to great lengths to explain himself and at the deepest level stuck to his guns to the bitter end. The assumption that caused all the opprobrium, the assumption that no one would accept, was his insistence on seeing, besides the evil, something positive in a regime that perpetrated the Holocaust.

More than twenty years after the war, my late friend Valentine Telegdi, the brilliant experimental physicist, ran into Heisenberg. The two had met many years earlier, when Telegdi was Paul Scherrer's student in Zurich. Telegdi complimented the grand old man on his memoir *Der Teil und das Ganze* (*The Part and the Whole*).

"In particular the part where you speak about your youth is very valuable for future generations. Of course, the part where you speak about the war is of a more subjective nature."

To this polite provocation, Heisenberg replied, "You know Herr Telegdi, the question of the war is a very complicated one. It is like the Arab–Israeli conflict now, nobody can say who is right." Valentine found this analogy so outrageous that he there and then politely took his leave and left a stunned Heisenberg standing alone in the big hall in which they had met.

Besides these moral issues, there is also a physics issue. As the Third Reich's most distinguished physicist, Heisenberg was put in charge of the Nazi attempt to produce a nuclear reactor and/or an A-bomb. This project turned into an unmitigated fiasco, not only in that Germany did not command the necessary technological resources, but also in that the science was flawed. After the war these scientific blunders, especially when contrasted with the immensely successful Manhattan project in the US, were a major embarrassment for Heisenberg. The main source on these matters is to be found in the famous transcripts of bugged conversations at secluded Farm Hall, outside

Cambridge, where Heisenberg and nine other German physicists and chemists were detained after the war. Heisenberg is overheard giving Otto Hahn, the discoverer of nuclear fission (the actual interpretation of Hahn's discovery as fission is due to Lise Meitner), a wildly incorrect estimate of the critical mass needed for a chain reaction. Yet days later Heisenberg gave his colleagues a clear and correct presentation of the physics of an A-bomb. Could he have known it all along and, by withholding his knowledge, sabotaged the German A-bomb effort? If he did, he acquires a heroic stature and the scorn heaped on him for scientific reasons is thoroughly misplaced. Or, could it be that brilliant as he was, he worked out the physics while in detention after the *fact* of Hiroshima? In that case, scientific blunder charges compound the moral charges. We will never know which of these alternatives corresponds to the truth. The bottom line is that for whatever reasons the Germans were unable to build nuclear weapons and in his youth Heisenberg made some of the greatest contributions ever to the science of physics. Let's be grateful for both.

Heisenberg's postwar work on the unified theory occurred at the same time he was sparing no effort to justify himself to his friends and ultimately to himself, and this effort may have exacted a great price. Before the war he had been at the center of things, everybody vying for his attention. He was also a great teacher. The list of his prewar students reads like a theoretical *Who's Who*: Nobel prize laureate Felix Bloch; the great physicist Edward Teller, though best known for his role in building the hydrogen bomb; Sir Rudolf Peierls, the great solid-state and nuclear theorist, who together with Otto Frisch was the first to get the correct value for that critical mass; the Romanian solid-state theorist Şerban Ţiţeica, who after writing a brilliant thesis under Heisenberg's guidance, returned to his homeland, physics-wise never to be heard from again; and a host of others. In the immediate postwar years he produced two brilliant physicists, the late Kurt Symanzik, and Reinhard Oehme, my University of Chicago colleague. But once the unified theory work got underway, Heisenberg became ever more isolated and surrounded himself with people awed by him, people to whom anything he did seemed to make sense. For a very brief period he struck up a collaboration with Wolfgang Pauli. A preprint (a prepublication copy) of a paper by Heisenberg and Pauli was circulated, but during a visit to America Pauli came to his senses and bailed out. None of the published work ever listed him as an author.

In 1958 things came to a head between the two. At a major international conference in Geneva during a session chaired by Pauli, Heisenberg presented his theory, ignoring all of Pauli's earlier criticisms. The two men began to argue, Pauli both using and abusing his prerogatives as chairman. Two giants who in happier days had collaborated on a major masterpiece now tore into each other, and about what? About a theory without any merit. This was still before my time, but I have heard numerous accounts of this session. In a sense it was a kind of Twilight of the Gods, a changing of the guard in theoretical physics. Pauli died shortly after this session: cancer of the pancreas kills very fast. Heisenberg ceased to be taken seriously.

3

A True Aristocrat

What most people who witnessed the Heisenberg–Pauli confrontation seem to remember is not the clash of the titans, but the comments made from the floor by Ernst C. G. Stueckelberg de Breidenbach, the major Swiss theoretical physicist. His remarks are recorded in the proceedings, but they are heavily edited. Apparently he had an episode of a recurring psychiatric condition, perhaps induced by the drama of the situation, and he talked as much about all kinds of characters hounding him during the break and hiding during the session, as about physics. It was quite eerie they say.

After the session Stueckelberg went back to his office at the University of Geneva on the banks of the Arve, grabbed his gun, and got himself arrested on the streets of Geneva brandishing it. You don't get far in Geneva with that kind of thing. They institutionalized him and from the institution he wrote a disjointed letter to the rector of the University resigning his professorship. In an inexcusable display of hypocrisy (Stueckelberg's psychiatric condition was hardly news by then) and snobbery his resignation was accepted and the University offered Stueckelberg's chair to all kinds of luminaries. To the everlasting credit of my profession, in spite of all the enticements, none of those contacted was willing to accept a position from which a master like Stueckelberg had for all practical purposes been fired. In the end, the Swiss physicist Joseph Maria Jauch accepted the position as head of theory in Geneva, but only on condition that Stueckelberg be reinstated with full back pay. The result was a first for the University of Geneva: two full professors of theoretical physics.

My first postdoctoral job had been in Geneva and I got a lot out of Stueckelberg's presence there. He was a tall man of striking appearance, displaying the easygoing elegance of a high aristocrat. His family was related to the German imperial family and in the eighteenth century, his ancestors had built a castle in Breidenstein zu Biedenkopf near Marburg, on the place where their fourteenth century fortress, by then in disrepair, had stood. Stueckelberg spent his summers there. He would pack a big trunk full of books and notes and be picked up by the local coachman at the railway station. His family had been the feudal lords of the region, and among his many titles, he once told me, was "protector of the Jews." Stueckelberg came to work wearing a hunter's outfit, gray knickerbockers and a green jacket, a pipe perpetually lodged in his mouth. He was inseparable from his dog, a small reddish-brown creature — a Swiss retriever if there is such a thing. The dog adoringly watched his master while dulling his keen sense of smell with the master's extremely aromatic pipe tobacco smoke. Besides his expertise in fine tobaccos, this dog was reputed to be quite a physicist. He came to all the seminars and Stueckelberg assured us that the dog was well behaved and only barked if he detected a mistake in the speaker's argument. It happened once, and indeed Stueckelberg, who until that moment had seemed rather uninterested in what the speaker had to say, asked question upon impatient question to spot the reason for the Swiss retriever's dissatisfaction. The only reasonable explanation I have ever heard for this dog's behavior is that far from being the great expert in theoretical physics his master had him be, the dog was the world's leading expert on its master's moods. The dog had not caught the speaker's mistake, he had simply detected that Stueckelberg was getting nervous because of it. The dog had even been granted right of access to CERN (The European Organization for Nuclear Research) by the director general of this great laboratory. He is reputed to have been the only certifiable son of a bitch ever to attend a seminar there.

Stueckelberg had made some very deep and fundamental discoveries in his time. He realized that if a particle of a certain type, the so-called π-meson, some 280 times heavier than an electron, were to exist, then one could readily explain why the nuclear force which holds nuclei together only acts at very short distances — some hundred thousand times shorter than even the size of an atom, as opposed to say the electric or gravitational forces whose range extends over macroscopic distances. The prediction of the π-meson — later experimentally discovered — is now generally attributed to Hideki Yukawa, who received the Nobel Prize for his work. Stueckelberg let Pauli talk him out of it. Pauli had told him it was the most stupid thing he had ever heard and warned him that his respect for Stueckelberg would vanish, were he to publish it.

Brilliant, very thorough, very conservative, and very critical, Pauli had repeatedly talked others out of major discoveries as well. The first physicist to discover the spin of the electron as the natural explanation of some experimental results in atomic physics was the twenty-one year old German Ralph Kronig, but Pauli told him it was all nonsense and convinced him not to publish. It was left to two young Dutch physicists, Sam Goudsmit and George Uhlenbeck, to rediscover the electron's spin. They also hesitated, but their great teacher Paul Ehrenfest sent their paper to be published, while they were out of town. They were never awarded the Nobel Prize for their discovery, partly because of the Nobel committee's qualms about whether this would be fair to Kronig. Depending on how you count, two or three others were deprived of Nobel Prizes because of Pauli as well. In 1956 when the Chinese-American physicists T. D. Lee and C. N. (Frank) Yang proposed that the laws of nature have a handedness — that they would not be the same were we to interchange left and right — Pauli went at it yet again. The way he put it, if God can't tell left from right with the strong, electromagnetic and gravitational forces, why would He have to tell them apart with the only remaining fundamental force, the weak one? This time, however, Pauli had met his match. Lee and Yang were so sure the laws of nature had a handedness, they did not budge, rather they challenged Pauli to a bet. Pauli did not take the bet. When he learned of the experiments confirming the Lee–Yang proposal, Pauli wrote to his former assistant Vicki Weisskopf, "It is good that I did not make a bet. It would have resulted in a heavy loss of money (which I cannot afford); I did make a fool of myself however (which I think I can afford to do) — incidentally, only in letters or orally and not in anything that was printed. But the others now have the right to laugh at me."

Beyond the meson, Stueckelberg had made some other key discoveries as well. The law of baryon number conservation largely responsible for our stable existence, as opposed to our quick decay into God knows what, was discovered by Stueckelberg. During the war he made major progress in constructing a consistent quantum electrodynamics, and in particular proposed the graphic representation we now call Feynman diagrams. Stueckelberg's theory of vector mesons is still being studied seventy years after he proposed it. It presents the first and simplest case of a phenomenon studied in its full generality by François Englert, Bob Brout, Peter Higgs and Phil Anderson in the sixties. Stueckelberg and his student André Petermann were the first to introduce the concept of renormalization group, which plays a central role in the modern quantum theory of fields. I am singing here the extensive praises of E. C. G. Stueckelberg de Breidenbach, because I think of him as a true, though widely under-appreciated genius.

I loved talking to him about physics. He had a knack for asking the right question — after talking to him I would look at what I was doing from a new angle. But it was by no means easy to talk to him. He demanded that each symbol you wrote on the blackboard be decorated by hats, hačeks (i.e. upside down hats), twiddles or bars according to whether the physical quantity it represented was even or odd under space reflection and under time reversal, even though for the matter under discussion neither space reflection nor time reversal may have been of any relevance whatsoever. Having to track these irrelevancies wasted a lot of time and concentration. Once I became aware that I had inadvertently made a mistake — I had put a hat where a haček was called for — but Stueckelberg didn't notice. So, I went on deliberately mixing up all these irrelevant ornaments, and he was oblivious to it. I then understood it was not that he needed to track all that irrelevant baggage. He just couldn't stomach "naked" letters, be they from the Latin or the Greek alphabet; he liked his letters neatly dressed. From then on I kept sprinkling random hats and twiddles and whatnots on my symbols, but now we could both concentrate on the matter at hand and not on these silly ornaments. Unfortunately, Stueckelberg's papers are also so adorned and to the uninitiated reader they look forbidding. Moreover, they are mostly written in French, hardly the *lingua franca* it had once been, and published in an excellent and prestigious Swiss journal, which however does not quite enjoy mass circulation. Nevertheless, we are dealing with a major physicist, a genius *à sa façon.*

Why was Stueckelberg never awarded the Nobel Prize? He was nominated, but the Nobel rules have it that a prize cannot be shared by more than three persons. Feynman, Schwinger, and Tomonaga, who shared the prize for quantum electrodynamics, pushed the theory further than Stueckelberg had, and though he had been first, he was the one dropped.

In a sense, Stueckelberg's discoveries provide the quintessential illustration of the principles of V. I. Arnold and Michael Berry:

Arnold Principle: If a concept, an idea, or a result bears a name, then this name is not that of its discoverer.

Berry Principle: The Arnold Principle applies to itself.

The Conscience of Physics

Now mediator bringing his friends to their senses, now merciless critic of the hopeless dead end, now spoilsport who discouraged many a major discoverer, now brilliant discoverer himself, Wolfgang Pauli hovers over his contemporaries as a kind of conservative and thoroughly honest supreme judge, an inquisitor defending physics. His colleagues dubbed him "the conscience of physics."

A few years before Pauli's birth, his Jewish father (born Pascheles), had converted to Catholicism, the religion practiced by his half-Jewish wife. The boy was baptized with the great physicist and positivist Ernst Mach as his godfather. He was a child prodigy given leave from attending high school classes so he could go to the University of Vienna to ... *teach* a course. Wolfgang Pauli Sr. was a professor at the University, and accordingly, for the early part of his career, Pauli signed his papers as Wolfgang Pauli Jr. His exclusion principle, for which he received the Nobel Prize, is essential for understanding the electron and the quark alike. It states that no two electrons can occupy the same state and it provides the explanation for the periodic table of the elements. Pauli proved the two most general theorems (the so-called spin-statistics and CPT theorems) without which relativistic quantum theory is unthinkable. Pauli postulated the existence of a new particle, the neutrino, to explain why no energy is lost in the process of radioactive beta-decay and that, on the contrary, the law of energy conservation applies in that decay process. This neutrino was then discovered and turned out to be one of the fundamental forms of matter. Along with Heisenberg, whom he met in Munich when they were both students of Arnold Sommerfeld, and with Dirac, Pauli is one of the founders of quantum field theory. His early work on Pauli paramagnetism largely launched modern solid state physics. His volume on General Relativity, written at the age of 21, earned him Einstein's admiration. In short, though not quite a household name like Einstein and Heisenberg, Pauli *is* one of the central figures in twentieth-century physics.

In 1928 Pauli was professor of theoretical physics at the Federal Institute of Technology (ETH) in Zurich. He was not the easiest man to get along with. He would go to seminars and like a davening Jew rhythmically nod his approval of everything the speaker was saying, until that moment came, as it almost always did, when the speaker said something Pauli disagreed with, at which point intense shaking of his head replaced the nodding and most speakers fell apart. Vicki Weisskopf, Pauli's assistant in the mid-Thirties, devised a strategy to cope with Pauli's nodding-to-shaking transition. In the morning of the day he was scheduled to give a talk, he would go to Pauli's office and give Pauli a preview. Pauli would tell him it was sheer madness and would forcefully try to talk him out of it. Then at the seminar when the "dangerous stuff" came, instead of shaking his head, Pauli would continue to nod while every now and then muttering to himself, "I told him this is madness." Apparently, though successful, Weisskopf's strategy was not widely known. It is said that upon his death in

1958, Pauli arrived in heaven and was summoned to God's office. Pauli asked God how it all works, and God went to the heavenly blackboard and started giving Pauli a private lecture. Pauli nodded at the beginning, but some twenty minutes into the talk he started shaking his head. Completely unnerved, God imprudently mentioned this to a few of the archangels, who spread it throughout the heavens causing universal heavenly consternation. Ultimately this was even leaked to earth by some lesser angel no doubt — that's how even I heard of it.

Someone this merciless with his younger and older colleagues and even with the very highest authority could not be expected to be much easier going when it came to students. A student at Cambridge, Homi J. Bhabha, who later became one of the first internationally recognized Indian theoretical physicists, was sent by his adviser to work with Pauli in Zurich. The introduction letter said, "Pauli, you are the only one who can possibly make a physicist out of Mr. Bhabha" — hardly a very flattering thing to say. Pauli, who much respected the author of this letter, interpreted it to mean that this was a hopeless case. Accordingly, he refused to have any meaningful talk with Bhabha, convinced that it would be a total waste of time. Whenever he noticed Bhabha in the hall, he would shout at him past the many other students hurrying to their classes, "Mr. Bhabha, what nonsense are we working on today?" The young man was despondent. It is, though, in Zurich that he wrote his most famous paper on electron–positron scattering, or "Bhabha scattering" as it is now called. When he tried to show the paper to the great man, Pauli responded, "If *you* did this, I am not interested." At wit's end, Bhabha contacted the other senior theoretical physicist in town, Gregor Wentzel at the University of Zurich. Familiar with his good friend Pauli's quirks, Wentzel took the paper and read it. He immediately realized its importance and assured its young author that he would convince Pauli of its merit.

At first Wentzel's attempt to explain Bhabha's work to Pauli was met with the expected, "If Mr. Bhabha did it, I am not interested," but Gregor was prepared for this. "For just a moment," he suggested, "imagine that I had done it." Pauli was willing to listen. Bhabha was forever indebted to Wentzel.

It wasn't only Bhabha — Pauli could intimidate other students as well. He would meet a group of students in the hall and ask them their names and ages. After they all introduced themselves, he would look dreamily ahead, saying: "Yes, at your age I was already very famous" and walk on. From a former child prodigy, maybe this isn't all that outrageous. But at heart, perhaps, Pauli was less secure than he let on.

There is a certain pattern to Pauli's behavior. He talked Kronig out of the idea of electron spin. Then Goudsmit and Uhlenbeck published the idea and it was rapidly accepted. The formalism to deal with a spinning nonrelativistic electron was then constructed by none other than Pauli himself. To this day we use that formalism and the famous Pauli matrices he introduced in that paper. He talked Stueckelberg out of mesons. Then Yukawa published his work about the meson and Pauli spent the war years in the U.S. working on meson theory. He arrived in the States rather suddenly, for he became acutely worried at the pro-Hitler sympathies of the Swiss. If that country were taken over by the Nazis, like his native Austria, then he, "under German law 75 percent Jewish," would be in for trouble. Even as is, the Swiss rejected Pauli's two naturalization attempts, which meant that in the wake of Austria's *Anschluss* he was a citizen of the Reich, a rather precarious status. In the States he ended up at the Institute for Advanced Study in Princeton, but first he spent a few months at Purdue University in West-Lafayette Indiana. An old acquaintance he ran into there asked him how he liked Indiana. "Just like Switzerland minus the Alps," Pauli replied. An apt characterization, if you think about it.

Pauli's personal life was not devoid of tragedy. In 1927, in the wake of some serious marital troubles, his beloved mother took her own life. In 1928 his father remarried and Pauli now had what he called an "evil stepmother." In 1929 Pauli left the Catholic Church, and soon thereafter entered a disastrous marriage with a cabaret performer, which ended in divorce less than a year later. She left him for a chemist named Goldfinger, of all things. Pauli was devastated and his father, a resident of Freudian Vienna, suggested psychoanalysis. Pauli ended up going to Carl Gustav Jung. The relationship with Jung lasted over the years and led to the book *The Interpretation of Nature and the Psyche* by Jung *and* Pauli in which Jung, with the approval of the Conscience of Physics, put forward the preposterous idea of some deep connection between the statistical nature of quantum physics and what amounts to extrasensory perception. Pauli's contribution to the volume concerns archetypes in the work of German astronomer Johannes Kepler. It is the kind of paper one expects from a graduate student, and not a particularly resourceful one at that. Here is a man with the highest standards when it comes to physics, going along with outrageous misapplications of the same physics when it comes to the stuff that dreams are made of.

Over many years Pauli supplied Jung with detailed descriptions of over one thousand of his dreams. To keep the sample as pure and unbiased as possible, Jung did not see Pauli himself; he only supervised the young psychiatrist seeing him. One cannot escape the impression that Pauli reveled in his role in this experiment and completely suspended his habitual disbelief. In the book *Psychology and Alchemy* Jung gives his take on Pauli's dreams and to this day, whether they know it or not, Jungian psychoanalysts have to read the dreams of one of the greatest physicists of the twentieth century as part of their training.

Remarkably enough, Jung and Pauli even took the so-called "Pauli effect" seriously. This effect amounted to the claim that whenever Pauli entered an experimental colleague's laboratory, something supposedly went seriously wrong: an expensive piece of equipment broke or burned out or otherwise malfunctioned. This very much intrigued Jung, who was very interested in the paranormal. Yet some experimentalists managed to come up with a simple precaution against the Pauli effect that clearly belies its paranormal nature. In a prominent place in the lab they installed extra large circuit-breakers not connected to anything. The moment Pauli entered the lab he would go right for one of these "Pauli switches" and would repeatedly switch them on and off and on and off without causing any damage. On the other hand, at a celebration of Pauli, his friends had prepared the room in which the banquet was to be held by installing a large chandelier rigged up to crash the very moment Pauli entered the room, a prank to "demonstrate" the Pauli effect. When Pauli made his entrance, the chandelier did *not* crash. The Rube Goldberg device which was to activate it, malfunctioned. Pauli effect?

It is sometimes said that people cannot live without some form of religion. Look at the Russian religious renaissance after seventy years of officially enforced atheism. Then look at Pauli, the baptized son of a converted Jew who leaves the Catholic Church only to have the aftermath of a painful divorce land him at the doorstep of a fascinating and charismatic medicine man dabbling pseudoscientifically in the occult. But then, divorces can play dirty tricks on the human brain — notice I refrain from using the ill-defined word soul. Even supremely rational and rigorous mathematicians are not exempt.

The Wizard of Pasadena

Unlike the other founders of modern quantum field theory, Richard Feynman, did not only formulate quantum electrodynamics as a workable theory, he first *reinvented* all of quantum theory from scratch. His sum over histories approach to quantum theory, equivalent to the earlier Heisenberg approach, is better suited for modern needs.

In classical physics, applicable in the familiar macroscopic world, point-like bodies clearly move along well-defined paths that record their history as it were. Determining these paths, or these histories, is the central issue in Newton's classical mechanics. But at the microscopic quantum level we can no longer speak of a particle moving from one given point to another along a *set* path, because there would be no experimental way to determine this path.

In Feynman's sum over histories approach this particle is viewed as having moved simultaneously along *all* conceivable paths from the one given point to the other, thus acquiring all conceivable histories. Feynman provides a formula for assigning to each history a characteristic number. Then he adds up these characteristic numbers of *all* histories and from the resulting sum reads off the probability of finding the particle at its given point of arrival at the time of arrival, if at the time of departure the particle was at its given point of departure. This is the end of the road. We no longer have calculable paths. If we had a calculable path, then knowing when and where the particle got started and when and where it arrived, we could figure out where it had been all along the way. Now, after summing over all histories, all we can predict is this probability of finding the particle at some point of arrival at the time of arrival, if we know when and where the particle got started. There is such a probability for any point accessible to the particle and the only thing we can be sure of is that the particle will be somewhere, so all these probabilities for it arriving at all these points have to add up to the probability of certainty, which equals one.

In retrospect, the method Feynman invented is even more important than the problem he solved with it. It took some time for this to sink in, and not until the Seventies did his approach start cornering the market. Everything Feynman ever did is marked by a simply devilish degree of ingenuity. The same is true of everything Einstein, Heisenberg and Gell-Mann did, but with them it is possible to see the wheels turning and to imagine that you could have done it the same way, if you were as smart and perspicacious as they were. But with Feynman, once you understand his idea, you cannot but admire what he did, but you won't see the wheels turning. Why did he do it that way? Your gut response is: sure, he has to be a smart aleck; he can't do it like everyone else. But sooner or later you realize, his is *the* way to do it, and how he does it is even more fun than what he is doing.

Feynman has a famous paper about the polaron in which he considered an electron in a crystal lattice made of electrically charged ions. Because of its electric charge, the electron distorts the lattice in its immediate neighborhood. When this electron moves, this lattice distortion moves along with it, as if a heavier particle, the polaron, consisting of the electron and of the lattice distortion, were moving. He once told me about this famous polaron paper. "One day, I was bored and went to the library. I took out a journal and started reading the paper of this guy Fröhlich in which he poses the polaron problem. He says it's very hard and very important, for once it's solved, we'll right away understand superconductivity [which at that point was not yet understood]. So I look at this problem and it was obvious to me that I could solve it. I went back to my office, solved the problem and then I sent Mr. Fröhlich a letter showing him my solution and asking him to let me know by reply mail how superconductivity works. Of course he couldn't!" Feynman laughed roguishly. "You know," he went on, "they all were shocked that I used my sum over histories, but to this day no one has been able to do it any other way."

Feynman's paper introducing the sum-over-histories approach to quantum theory is one of the most beautiful physics papers I have ever read. He invents rich new mathematics and a new way of looking at quantum physics. It is given to but very few physicists to invent truly significant new mathematics — in the twentieth century this can only be said about Paul Dirac, Richard Feynman, Edward Witten, and Pascual Jordan.

Feynman wrote a book on the new sum-over-histories (or "path integral," as it is often called) approach with A. R. Hibbs, a former student of his. He was not satisfied with this book, which seems to lack the excitement and magic of the papers on which it is based. He once told me, "You know Hibbs wrote the book and then asked me to put my name on it. I guess I shouldn't have."

Feynman wasn't all that protective of his former students. Another time, when I was visiting Caltech for a quarter, I wanted to know more about an unpublished but at that time relevant thesis of a former Feynman student. Feynman came to my office and treated me to a superb lecture on the subject. It was vintage Feynman. I loved it and he saw that. At the end he added, "You know it was really I who did all this." This was unnecessary.

The most remarkable and charismatic lecturer I've ever heard, Feynman used his whole body in a natural highly effective way to convey the material. He loved playing with his audience. Once at a workshop he lectured on two topics. Over lunch he confided in me. "Freund," he always called me by my last name, "I fooled them. In the lecture on partons, I made them believe that it was all child's play and that I was only waving my hands, but I have detailed calculations in my notes and I can support every claim I made with an unassailable calculation. Then that other stuff, I made it sound like it was God knows what kind of high theory. It's all hand-waving."

There were a number of workshops and conferences where we both lectured. In 1972 we both lectured in Mexico. I am sure of the year, because I remember reading in a Mexican newspaper about some break-in at some building complex called the Watergate. It was front-page news and I distinctly recall thinking, probably some third-rate burglary, but these Mexicans exaggerate everything. We stayed at the same hotel in the Zona Rosa, and for three evenings Feynman came to my room and asked me to lecture him about string theory, then still in its infancy. He would interrupt me with, "Freund, can I ask a stupid question?" and then ask the kind of question to which you would answer either, "That's exactly what I am coming to" or "Unfortunately no-one has figured that out as yet." After the fifth or sixth such question I said to him, "Look Feynman," maybe he called me Freund because he didn't want me to call him Dick, "You know very well that these questions are very smart." He smiled. I went on, "In fact this is by far the best class I have ever taught." His smile broadened. He was playing a game and enjoyed the fact that I was playing along.

In spite of all these games, Feynman was a deeply honest man. We were once lecturing at Fermilab, the big accelerator laboratory outside Chicago. During Feynman's lecture, Glennys Farrar pointed out that his claim that one could directly measure the electric charge carried by quarks was incorrect. He seemed sure of its correctness and dismissed Glennys' remark. She persisted, and to end the matter Feynman made a joke at her expense. Everybody had a good laugh, and the matter seemed closed. But after the talk Glennys, not one to give up easily, came to him and explained what she meant, she had already written a paper with Jon Rosner, my colleague-to-be, on the subject. Feynman got the point, but didn't believe it and said he would think it over. I was staying at the Fermilab hostel, as was Feynman and my friend Steve Ellis. I remember the three of us discussing the matter to death. The most desperate attempts were tried to salvage Feynman's claim, but it was becoming ever clearer that Glennys was right. Around midnight we gave up and went to our rooms. I remember Feynman's face when we said goodnight, I also seem to remember some unprintable words that were spoken.

The next day Feynman was scheduled to lecture on another topic. He could easily have avoided any reference to the previous day's confrontation. Instead, barely did he get going, when he brought the matter up. "Yesterday, Ms. Farrar remarked that my claim about measuring the quark charges was wrong. I made light of her remark, but after my talk she explained her point to me in detail. After thinking it over, I reached the conclusion that she is right and I was wrong, because of careless reasoning on my part."

I think everyone in the room was very impressed by his frankness. Feynman went way up in my esteem; he was clearly not only a genius but also a *mensch*. He could have said, "Ms. Farrar's point may be well taken under certain reasonable assumptions" or some other kind of wishy-washy mumbo-jumbo. He could have said, "Ms. Farrar is correct, but the claim in question is just a minor point," or some other defensive stuff. No, he openly admitted to what was, after all, not a major mistake. That takes courage, and Feynman had what it took.

Back to Maupertuis

A very remarkable consequence of Feynman's sum-over-histories approach to quantum physics is that if some "classical" phenomena are the macroscopic limit of what in reality are a large number of microscopic quantum phenomena, then the laws of nature governing these classical macroscopic phenomena must be of a very special kind; they must be what the mathematicians call *variational principles*. What this means is that there must exist a prescription which from every conceivable history of the system — whether or not this history respects the classical macroscopic laws of nature — allows us to calculate a quantity, which for the history mandated by these classical laws takes its least possible (more generally, extremal) value. So you can find the history mandated by the laws of macroscopic classical physics by minimizing this quantity, usually called the action. A classical system can only have a deeper quantum origin if its classical law takes the form of such a least action principle. Dirac already knew this. In the older classical physics it had long been known that the laws for all "fundamental" phenomena can be cast in such a least action form, but this was generally viewed as a mathematical curiosity, and not as the harbinger of a deeper microscopic description. It took Dirac and

Feynman to recognize it as such. You may ask how it was that people who had never heard of quantum physics even came up with such a least-action description for classical physics. The way they did is nothing short of miraculous.

It all started with a Frenchman, Pierre Louis Moreau de Maupertuis, who had made a name for himself with a 1736 expedition to Lapland during which he confirmed Newton's prediction about the shape of the earth as an oblate spheroid. A great admirer of Newton's by then well-entrenched ideas, Maupertuis asked himself whether the very existence of laws of nature was compatible with the existence of God (I am serious). He claimed the two were consistent and, even more, that one could prove the existence of God by studying the way this consistence is achieved. Let me briefly paraphrase Maupertuis' reasoning. First of all, there was nothing wrong with the Almighty simplifying his manifold tasks by setting down some laws. Why should He be distracted from important matters involving war and peace, kings and queens and what have you, by having to make sure that each pebble on the beach moves as it should, and that ships in heavily trafficked seas don't collide for no reason at all. Once one buys the idea that He imposes some laws of nature, one must ask what kind of laws one should expect from Him. Well, wherever you look in the Holy Scriptures you are being admonished to please Him, in order to make for a better world. A God this intent upon being pleased, reasoned Maupertuis, would no doubt set down laws which maximize His pleasure, or equivalently minimize His displeasure. If Newton's laws are compatible with the existence of God, one should be able to recast them in a form in which some quantity — whether you call it "God's displeasure" or "the action" — is minimized, and *voilà*, you've got your least action principle.

Maupertuis, no math whiz himself, managed to find that such a least action principle was indeed at work in

some simple cases. The idea of proving God's existence
along these lines made him immensely famous. Frederic
the Great, the Prussian King, a great admirer of all things
French — he promoted the use of the French language at
his court — imported M. de Maupertuis and made him
president of the Prussian Academy. Not surprisingly, the
other academicians scrambled to learn what their new
president was so famous for. Among them was one of the
greatest mathematicians of all time, the Swiss Leonhard
Euler, who gave Maupertuis' idea the mathematical treatment it rightfully deserved — and variational calculus, a
major branch of mathematics was born. In all fairness, the
Torinese mathematician Joseph-Louis Lagrange also had
an important hand in the creation of variational calculus,
as did Sir Isaac Newton, who had anonymously sent in the
solution of the first modern-era variational problem, that of
the brachistochrone (how's that for a name?), proposed by
Johann Bernoulli, a member of the legendary Bernoulli
family which gave the world no fewer than eight first-rate
mathematicians. Newton, who had retired from the sciences over a decade earlier, sent in a solution anonymously
because, in a very English fashion, he did not want to be
"dunned and teased by foreigners about mathematical
things." It didn't work though, Bernoulli claimed to have
"recognized the lion by his claws."

Of course, Maupertuis had not proved the existence
of God. At best he showed that the laws of Newton are not
incompatible with Maupertuis' own reading of the
Scriptures. Still, this whole line of thought was of great
interest in his day. After Newton, phenomena previously
thought to be ruled by the momentary wishes of one or
more deities were now becoming predictable. Solar eclipses
were no longer omens of celestial wrath, but perfectly predictable occurrences, the paths of celestial bodies could
now be calculated without any reference to a Divine Being.

So where were the new boundaries between the sacred and the profane to be drawn? The long process of redrawing boundaries had already begun with the trial of Galileo by the Inquisition, and in some sense is ongoing even in our day. At one point Pope Paul VI wished to bring at least the Galileo episode to a closure, without a loss of face for the Church. He appointed a commission made up of members of the Pontifical Academy in Rome to study the full details of the case with the goal of rehabilitating Galileo in mind. Even Galileo's canonization came up as a remote possibility: the Church would make amends to the great scientist whose requisite "heroic virtue" during his mistreatment by the Inquisition would qualify him for a saint's status, so that henceforth praising Galileo might mean praising the church he was a saint of. The Pope had the best intentions for settling the matter. On the commission he appointed Joseph Maria Jauch, E. C. G. Stueckelberg de Breidenbach's Geneva colleague. Jauch told me that it was very interesting work, one of the popes Galileo had dealt with having himself been an amateur astronomer who kept asking valid questions. The commission's work was progressing smoothly when Jauch's marriage suddenly ended in divorce. Was this interpreted in Rome as an ill omen? In any case, to this day poor Galileo is not recognized even as a martyr, let alone a saint with fully shining halo. Nevertheless, in 1981 Pope John Paul II appointed a new commission to study the Galileo case. This second commission handed in its final report in 1992, and on October 31 of that year, the Pope addressed the Pontifical Academy on the subject of Galileo. His discourse can be interpreted as a rehabilitation of the old rascal.

Were we to take Maupertuis at his word and believe that the classical path of an object is indeed the one that least displeases God, we might have to ask where this leaves Feynman and his sum-over-histories. According to

Feynman, at the quantum level of the microscopic world, all paths are traveled, there is no further distinction between godly paths and sinful ones, every path gets a chance. It is only in the classical limit that minimizing the displeasure of the Almighty is *the* thing to do. This brings up the oft-debated issue of why a God so strict about being pleased ends up tolerating Holocausts, scourges and epidemics that claim so many righteous lives. Could it be that God no longer listens to the French, but is Himself taken with Feynman? But I should stop musing this irresponsibly. What if a philosopher, a jurist, or one of those ethicists — I think that's what they call themselves — were to read this? God forbid!

The Language of God

In the nineteenth century, with such important worries as "when does life begin?" or "does a woman have the right to choose?" still in the distant future, that ever eager species, the "public intellectual," had to find something to worry about, so the worry *du jour* was: "If there existed a feral child, reared by wolves or bears in the wilderness, away from any human civilization, what language would he be speaking?" Most of these public intellectuals agreed that this purest of humans would be speaking God's own language, Hebrew (not old-English, as some might have guessed). Then, in 1828 the next best thing to such a noble savage showed up in Nürnberg and proved all the public intellectuals wrong.

Known as Kaspar Hauser, this lad neither spoke Hebrew nor any other language; he didn't speak at all. So, I will just assume that God's language is mathematics and rather than relying on observations of noble savages, I will just refer to the undeniable fact, emphasized already by Galileo, that the laws of nature, as we know them, are all expressed in this language. Since physics studies precisely these laws of nature, not surprisingly, an intimate connection between physics and mathematics is unavoidable.

We have seen how, on rare occasions, physicists can create rich new mathematics. Most physicists, however, go and raid the shelves of the mathematicians and show how existing mathematical structures are custom-made to describe nature. Think of Einstein who, in constructing his General Relativity, used the already existing geometry of spaces in which the concepts of distance and curvature are defined — a geometry discovered by Riemann in the nineteenth century. There exist a great many beautiful and important mathematical theories and mathematical objects. Occasionally, from this multitude mathematicians, presciently it seems, discover precisely those mathematical theories and objects that correspond to natural phenomena not even hinted at, let alone discovered in their time.

How is this possible? In the words of Eugene Wigner, how is one to account for the "unreasonable effectiveness of mathematics"? Let me try to answer this question by generalizing a picturesque description, originally due to Einstein, of how physicists go about their work.

Physicists come in two major breeds, theoretical and experimental. Picture all of natural phenomena as regions in a plane. The experimental physicists are wandering around this plane and exploring various regions in it. They report what they find. Now picture the theoreticians moving about in a higher plane parallel to the nature plane of the experimentalists. They look down at the experimentalists, no pun intended, and notice relations between the findings of experimentalists working in not too distant regions from one another. The theoreticians come up with a common explanation of all these results using tools from mathematics. By careful reasoning, they predict new consequences of their "theory" of these phenomena. The experimentalists now rush to the region in their nature plane suggested by the theoreticians and make the appropriate measurements. If the theoretician's prediction is borne out, then more theoretical work is done, new predictions are made, tested and so on, until one day something doesn't check out and then the theory is modified and a better theory, a better map of reality is constructed. Theories are living things, they are not eternal truths. They continue to hold in the region where they were first checked, but as larger and larger domains in the nature plane are covered, ever better and more general theories are constructed.

Now, to this two-plane picture of Einstein's, let me add above the theory plane a third, even higher plane parallel to the other two planes. In this plane live the mathematicians. They look down at the theoreticians, *certainly* no pun intended, and notice that what some theoreticians working in not-too-distant regions of the theory plane are doing can be encapsulated in a beautiful mathematical structure which they then explore as fully as possible. Their development of this branch of mathematics may then project down to regions in the theory plane quite far from those from where everything started, and these new regions in turn project down to the nature plane to regions which will be explored only a century later by the experimentalists. When these experiments get reported and their theory gets constructed, the theoreticians are stunned — the mathematics is all already in place. Unreasonable effectiveness? No, I would say. More indirectly, mathematics also has its origin in nature, and it should not surprise anyone if the mathematics of one century has consequences that find their correspondents in nature only in the next century.

Notice that both the experimentalists and the mathematicians live in the outer planes, whereas the theoreticians live in the plane sandwiched between them. This theoretical plane is not midway between the other two planes. Rather, it moves now closer to the plane of the experimentalists, now closer to the mathematicians' plane.

Right now theoretical physicists like me are so close to the mathematicians that it is hard to tell the theoretical physicists and the mathematicians apart. Historically, such epochs have coincided either with the births of major new theories or with the creation of the requisite tools for the next breakthrough. When a new theory is born, all three planes tend to come together. Not even the distance between the two outer planes is fixed.

There are many examples. In the days of Newton the distances between any two of the three planes was almost null. After all, Newton himself may have been the greatest experimentalist, the greatest theoretician, and the greatest mathematician who ever lived. There may be a very few others who come close to him in one of these fields, but not one soul who can match him in two, let alone in all three. The theories being born then were of course classical mechanics, Newton's theory of gravity, and his corpuscular optics.

When quantum mechanics was being born in the 1920s, the mathematicians Hermann Weyl and Johnny von Neumann (the latter well known as one of the fathers of the computer age and of game theory) were active participants. Not only can physicists produce new mathematics, mathematicians can also produce new physics. Hermann Weyl's idea of gauge invariance is one of the most fundamental in the theoretical physicist's arsenal. Had he but been alive in the late Seventies, he would certainly have been awarded a Nobel Prize for it. In the 1920s the theoreticians were also very close to the experimentalists — to James Franck, to Otto Stern, and to many others.

At present, as I have said, theoreticians are very close to mathematicians but, to be frank, quite remote from the experimentalists. The reason is that the energies to which we must accelerate particles in order for the effects of the "new physics," the physics of the theory now under construction, to set in, are expected even under the best of circumstances, to exceed the energies that can be reached with currently available accelerators. But the construction goes on, the blackboards on the theory plane are full. What does this mean? Are we or are we not amidst the birth pains of a major new theory?

Maybe another example is relevant here. During the period 1770 to 1830 it was also difficult to tell the mathematicians from the theoretical physicists. Euler, Poisson, D'Alembert, Fourier, Lagrange, Gauss, Laplace, Hamilton and Jacobi were all both leading theoreticians and leading mathematicians. What did they produce in physics? Lagrange, Hamilton, and Jacobi, for instance, gave mathematically brilliant novel approaches to Newton's mechanics. Now, a pure experimentalist of that period, someone like Ørsted or Ohm would probably have dismissed their work as pure mathematics, for to tackle any mechanics problem, the good old Newton equations were all you needed. Yet, can one even imagine how quantum mechanics would have been discovered a full century later, had Lagrange, Hamilton, and Jacobi not invented these powerful approaches? Sure, quantum mechanics can also be arrived at by starting directly from Newton's equations, but it is more something you do "for completeness' sake" than something that forces itself upon you. Maybe what is now being done in string theory is the latter day version of what Lagrange, Hamilton, and Jacobi did for mechanics, the true impact of which will not be felt for quite some time. Just as then, the mathematics of it is splendid.

Maybe this point can be made more easily by an analogy. When the telephone was invented, its use spread quickly and copper wire telephone lines were being laid all over the world. Then at some point the telephone companies replaced all these lines with optical fiber. Strictly speaking, just to call auntie Alice and wish her well on her birthday, you'd say good old copper lines can do the job, who needs all this optical fiber? But then can you even remotely imagine the information age *without* optical fiber?

All this raises the important question as to whether today's string theory is mathematics or physics. In string theory, the problem of unifying all forces and all known forms of matter with each other has been expanded to include the determination of the attributes of space and time, to the point at which we can "predict" the number of space dimensions and the shape of space. Instead of various types of point particles being the elementary forms of matter (like, say, the electron), and the carriers of force (like, say, the photon, the carrier of the electromagnetic force), in string theory they all are but different excitations of a more fundamental *extended* object, the *string*. This string obeys a number of laws of a rare mathematical beauty, and just as a usual string can produce various pitches depending on how we excite it, how we pluck it, or how we bow it, so this fundamental string also has excitations and among these we find all the familiar particles (the electron, the photon, and so on) and a host of predicted new excitations necessary for the mathematical consistency of the theory. The problem is that nobody has yet seen any of these new phenomena predicted by the string and in view of the possibly extremely high energies at which these will show up, we may for years to come, have to look for and be satisfied with what in jurisprudence is called circumstantial evidence. Then, as I just said, there is the possibility that the very beautiful mathematics dictated by the string, may by itself be the most important result of string theory, and is meant to be used in the future the same way as the mathematics of Hamilton, Lagrange and Jacobi was used when quantum mechanics was discovered.

8

Emmy Noether and the Urge for the Abstract

Nowadays mathematicians, the speakers of God's language, are spread out around the world; theirs is a democratic republic of genius. They have numerous regional capitals: Paris, the Cambridges, Princeton, New Haven, New York, Chicago, Berkeley, Kyoto, ... It hasn't always been this way. Before the collapse of the Soviet Union, Moscow would have clearly made this list, and further back, before Hitler, mathematics was not even a republic, but a principality with the princes residing in Göttingen, Lower Saxony. This situation dates back all the way to Carl Friedrich Gauss, the universally recognized *princeps mathematicorum*, prince of the mathematicians, in the first half of the nineteenth century, and runs through a line of genius: Bernhard Riemann, the Mozart of mathematics, the great number theorist J. P. G. Lejeune Dirichlet, Felix Klein of Erlangen Program fame, and finally the new Gauss: David Hilbert. At the 1900 International Congress of Mathematicians in Paris, Hilbert set the tone for twentieth–century mathematics by listing twenty-three problems, the solution of which he found central to its progress. It is nothing short of miraculous how clearly Hilbert saw what the mathematics of the dawning century would be about.

In 1915 Hilbert wrote one of the fundamental papers in general relativity, and this led him to a certain problem in invariant theory. To solve this problem he and Klein brought to Göttingen Emmy Noether, a woman, a highly unusual step in those days. She solved this problem by proving a theorem which provides the general connection between symmetries of the laws of nature and observed conservation laws. By way of an example, the laws of nature exhibit a symmetry in that they do not change in time — the laws are the same today as they were yesterday or billions of years ago when the distant quasars emitted their light signals which reach us today. This is experimentally ascertainable, for we recognize all the lines in the spectrum of this radiation using *today's* atomic physics, even though this spectrum was determined by the atomic physics valid at the moment of emission of this radiation very long ago. We express this property of the laws of nature, of not changing in time, by saying that the laws of nature exhibit *symmetry* under time translations. As a consequence of Noether's theorem, this symmetry yields a conservation law for the quantity called energy. This conservation law can also be tested in great detail. It says in particular that if I hit a particle with another particle, then no matter what processes should be called forth by their collision, whether just the two particles go their ways or some new particles are being produced, the total energy of all the outgoing particles must equal the total energy of the two incoming ones. This can be checked by measurements because Emmy Noether gives us a prescription of finding the energies involved. Einstein himself was very impressed by this theorem, but Emmy Noether later dismissed it as a bagatelle from her youth.

After this theorem, she returned to work in *purest* mathematics. Under her leadership the very style of mathematics was radically changed. Mathematicians began to keep an ever greater distance from the concrete origins of the problems under study, and generality became so highly prized that the structures studied by mathematicians became ever more abstract, even by the standards of mathematics. The reward was that in this purest abstract form new unexpected connections started emerging. By making dry algebra abstract, these abstract algebraic structures would be found to underlie much of topology, a subject which deals in juicier objects, — surfaces and their higher dimensional generalizations, and the way these can be mapped continuously or smoothly into each other. This way topology aims at understanding all possible truly distinct shapes such a generalized surface can have. Abstract algebra and modern algebraic topology, now subjects at the very heart of mathematics, got their initial start from Emmy Noether's vision.

Even under the best of circumstances, mathematics is already quite abstract. Increasing its degree of abstractness may be just a natural urge of its practitioners, so, perhaps this development was bound to come sooner or later. What is extraordinary is that it came in the wake of World War I, a historical event, not a scientific one. It could have happened earlier, it could have happened later, but the abstract became dominant in mathematics at about the same time it also made its appearance in painting (Kandinsky, Malevich, Mondrian), in sculpture (Brâncuşi, Arp), in music (Schoenberg, Berg, Webern), in architecture (Mies, Gropius), and in literature (Joyce, Stein). It is perhaps less remarkable that a move toward the abstract happened more or less simultaneously in all these other fields, for in painting, sculpture, music, literature and architecture, the artists mostly knew and influenced each other. An extensive correspondence between Kandinsky and Schoenberg on these matters has been preserved.

But Göttingen was a provincial town, and it is hard to believe that Emmy Noether cared all that much for the arts. So how is one to explain the fact that mathematics too veered toward the abstract at the very same time? Somewhat tautologically one could invoke that ill-defined concept *Zeitgeist*. Maybe after the Great War everybody wanted to get away from the ugly realities and going abstract was the way to accomplish this. Maybe people wanted to find the true causes of the disaster that had befallen them rather than hover at the surface of things, where they would be distracted by curlicues, by irrelevant details. So they decided to proceed to basics, to the abstract underbelly of mathematical and of artistic truth.

Physics as well became more abstract. Instead of planets moving around the sun or vehicles moving on the road, ever since Bohr's work we had electrons moving around the nuclei in an atom so small that there could be no way of observing the electron's path. Strictly speaking there is no such path; that was the clue used by Heisenberg in the construction of Quantum Mechanics and by Feynman in taking the *sum over* paths. We could still test the predictions of this quantum mechanics, but less intuitively than in its classical counterpart. You can follow Jupiter moving around the sun through a telescope but you cannot peer inside an atom to closely follow one of its electrons.

The possibility remains though, that this simultaneous move to the abstract in the sciences and in the arts was a one-time occurrence, an accident, a mere coincidence. I strongly doubt that, because the return from the abstract to the more congenial representational in painting and sculpture, to tonality in music, to the postmodern in architecture and literature *and* to the more specific in mathematics, all occur, again, more or less at the same time in the Sixties and Seventies, during the Cold War. In physics it was not so much a return to the old intuition as it was a development of a new intuition for that which yesterday was feared as being too abstract. Perhaps, when bombarded with the "abstract" ideological slogans of the opposing Cold War powers, mathematicians felt a need to return to the specific, even if this implied some loss of lofty generality. After all, physicists, mathematicians and artists are all members of the *same* society and as such subject to its moods. Had Maupertuis witnessed all this, he might have claimed to have found the meaning of the elusive *Zeitgeist*. He wouldn't have been any less justified than in his claim to have proved the existence of God.

At the human level the story of Emmy Noether, the *éminence grise* of abstract mathematics, is also remarkable. The daughter of an illustrious mathematician, the young Emmy, though educated at a fashionable school for girls, did the unfashionable thing for a girl and went on with her studies to obtain a doctorate in mathematics with Paul Albert Gordan, *the* expert in invariant theory. She was the only doctoral student Gordan ever had, but what a student! It was Gordan who got her to Göttingen, but that is as far as his influence reached. The University of Göttingen could not be expected to outright *hire* a woman. You start by hiring a woman and God knows what outrage comes next. So Emmy Noether was given the job of an unpaid research assistant. The money made no difference to her, the Noether family was independently wealthy. But from a few years after her arrival to Hitler's rise to power, the mathematical life in Göttingen to a large extent revolved around her. She had all the best students; visitors from far away places came to receive her advice; she taught extremely influential courses. She may not have been a great lecturer, but she had great things to say.

During World War I, Hilbert took it upon himself to champion Emmy Noether in front of the University senate, suggesting an appointment of her as assistant professor. The professors of humanities were outraged by the fact that a woman was even being considered. Hilbert famously quipped: "I don't see why the gender of the candidate is relevant. After all, our department does not run any public baths." Yet it was not to be. After the war she was finally "appointed" as an *unpaid* associate professor. This didn't disturb Emmy Noether much. What did disturb her was that she was not listed as an editor on the cover of *Mathematische Annalen*, arguably the world's leading mathematical journal at the time. In practice she performed many of the duties of an editor-in-chief, but you had to be a man to get your name on the cover. Her papers, which appeared *between* the covers of that journal, rapidly became classics, and her standing in the mathematical community rose to great heights.

When Hilbert retired in 1930, he personally requested that Hermann Weyl, his most distinguished pupil, be appointed as his successor, and in matters of princely succession there are no ifs, ands, or buts. Weyl came to Göttingen, but later claimed that he felt ill at ease with everything centered on Emmy Noether, still a lowly unpaid associate professor. She tried to fit in, to be one of the boys — so much so that the others used on her not the appropriate German feminine article, as in *die* Noether, but rather the masculine one, as in *der* Noether, or maliciously behind her back sometimes the insulting neutral one, *das* Noether. In his moving Noether obituary, Weyl, with Wagnerian zeal mentioned that "essential aspects of human life remained undeveloped in her, among them the erotic, which, if we are to believe the poets, is for many of us the source of emotions, raptures, desires, and sorrows, and conflicts." As to her appearance, he noted "no one could contend that the Graces had stood by her cradle." But he also sees in her a human being whose heart "knew no malice," who did not believe in evil. Above all, he sees in her a true mathematical genius.

In 1933 Hitler came to power. Emmy Noether was Jewish, and the Jews at Göttingen as everywhere else in Germany were being purged. The legendary analytic number theory course of Edmund Landau was cancelled in the wake of a student boycott led by Oswald Teichmüller. This man went to Landau and told him to his face, "We want to learn Aryan mathematics, not the Jewish kind." Richard Courant was forced out, came to the States and, a great organizer that he was, put New York University on the map. Emmy Noether went about teaching her course, and Jewish as it may have been, even the Nazi boys could not dispense with it. Teichmüller, Ernst Witt, and other brilliant students were in attendance, Witt in full Nazi regalia — politics is not mathematics after all, you don't have to be consistent. After forcing old Landau out of the classroom because he wanted his maths Aryan, Teichmüller, stuck on a problem, did not find it hypocritical to approach Emmy Noether for help, which she generously gave.

She had to leave, though, and in 1933 she came to the States. The flood of mathematicians from Germany, Jews and Gentiles alike, was of such proportions that jobs were hard to find. Weyl, a Gentile who also had to leave on account of his Jewish wife, was immediately appointed at the Institute for Advanced Study in Princeton, but Emmy Noether had no such luck. The job she was offered was not with a research university, but with a college, a girls' college at that: Bryn Mawr, a fine school on the outskirts of Philadelphia, but hardly the school where you expect to find an Emmy Noether. She accepted without much ado, because she would be close to Princeton, which was fast becoming the new world center for mathematics. At Bryn Mawr, in her inimitable fashion she started out Olga Taussky on her career as the first quite significant American woman mathematician. For the summer of 1934, she returned to Germany, not least to make sure that the talented Nazis she had left there were keeping up the good work in mathematics, never mind the mustachioed fellow. She came back to the States and in the spring of the following year, shortly after her fifty-third birthday, she died of cancer. The *New York Times* printed a short obituary as it always did when a Bryn Mawr teacher died, but shortly thereafter they printed a long letter to the editor pointing out that Emmy Noether had not only been a teacher at a girls' college but the greatest woman mathematician of all time. The letter was signed: Albert Einstein.

9

Oswald Teichmüller and Nazi Science

In Durham Cathedral I came upon the grave of Saint Cuthbert — quaint name — where I noticed, lay also the head of the Northumbrian warrior king Saint Oswald. Warrior king! No wonder the name Oswald has always filled me with an ominous premonition of evil. Osvaldo Valenti, an actor and a dyed-in-the-wool-fascist, starred as the evil-doer in many Italian movies during World War II when I was in grammar school. Towards the end of the war the partisans shot him and he disappeared from the screens, but in my heart he left an everlasting weariness of anyone named Oswald. I know it's irrational, and there must be good Oswalds, yet the ones that easily come to mind — Sir Oswald Mosley, the British prewar fascist; Oswald Spengler with his *Decline of the West*; Lee Harvey Oswald, the presidential assassin — only thicken the gloom. Beyond these I can think of only two others, Thorstein Veblen's nephew Oswald, the Princeton geometer whose personality I know next to nothing about and Oswald Teichmüller, who, as you might guess after what I wrote in the previous chapter, further strengthened my anti-Oswaldian prejudice.

Teichmüller was both a mathematical genius and a fanatic. In the end it cost him his life. The student who asked Emmy Noether for advice was then only a minor inconsistency in this man's life story. Teichmüller's boycott of Landau, idiotic though it may have been, at least fits in neatly with what this young man firmly believed. Let me be as detached as I can. Teichmüller bought Hitler's line about the mortal danger the Jews posed to German culture lock, stock and barrel. In mathematics he was under the spell of the ideas of the Berlin mathematician Ludwig Bieberbach. Author of some truly beautiful textbooks — I myself very much enjoyed two of them in my undergraduate studies — Bieberbach is remembered nowadays for a conjecture which took almost seventy years to prove. Although he has written some famous papers with his "friend" Issai Schur, a Jew, about ideas of Hermann Minkowski, another Jew, he found it expedient to champion the anti-Semitic laws of the Hitler regime.

Bieberbach went overboard with proclamations about an authentically German "synthetic" style in mathematics, which reflected the Germans' deep intuition, as opposed to the "analytic" style of the inferior Jews and, while he was at it, of the inferior French. These proclamations remind me quite strongly of Richard Wagner's essay *The Jews in Music* in which he rants against what he perceived as the Jews' lack of passion and their exploration of the irrelevant and formal without the daring reserved for the "German Masters." For Wagner it was part envy of others' success, part frustration, and part outright dishonesty. For Bieberbach it was one hundred percent opportunism. With the Jews out of the way, he saw his own role in German mathematics considerably enhanced. He started *Deutsche Mathematik*, a journal in which he urged his Aryan contributors to forgo the sweat of analysis and parade their Germanic intuition. In Oswald Teichmüller he found a willing taker. This young man's prophetic papers on what are now called Teichmüller spaces were mostly published in *Deutsche Mathematik* and some other *very* German journals in which a wanton disregard for common rigor in favor of that vaunted Germanic intuition was encouraged. The result was that for a long time Teichmüller was not taken seriously, at least not till the Fifties, when in the hands of the Finn Lars Ahlfors at Harvard and the Latvian Jew Lipman Bers at NYU his results were *really* proved with the customary clarity and rigor. Bers was fully aware of the irony of a Jewish refugee from the Nazis doing the work needed to vindicate the spectacular mathematical ideas of a rabid Nazi. He even went so far as to quote Plutarch on the matter: "It does not follow of necessity, that if the work delights you with its grace, the one who wrought it is worthy of your esteem." An apt quotation, it applies equally well to Wagner and alas to so many others.

As to Teichmüller, Bierberbach brought him to Berlin in the mid–thirties and he went about his discoveries there, though to say he went happily about them would be exaggerating. A famous story tells of the fire in the mathematics seminar room at the University of Berlin. Someone had set fire to the room and they couldn't blame it on the Jews; there were none left. Under the circumstances an investigation was made and all signs pointed to Teichmüller as the arsonist. He was called to the rector's office and asked, "Dr. Teichmüller, why did you set fire to the seminar room?" After a short pause he is reputed to have answered, "That is a good question." If this story is not apocryphal then it sounds like the fellow was outright out of his mind. But in his own way, he didn't lack a certain consistency. During the war, as a professor at the University of Berlin, he could easily have obtained a deferment, but instead, he volunteered for the front and got himself killed in Ukraine, and for all we know, he may have been shot by a Latvian Jew. He believed in his Führer, he strove for his Führer, he died for his Führer.

Not Mr. Bieberbach. He was left over after the war, but was totally ostracized by the mathematics community. Well, not quite totally. When I was a graduate student at the University of Vienna in 1960, I saw an announcement of a talk by Bieberbach. Having admired his textbooks so much and unaware of his Nazi past, I went to his talk just to see the man, for the subject held no particular interest for me. I remember a tall and somewhat overbearing old gentleman talking to the four or five of us. The others seemed not to be any more interested than I was. Bieberbach had a disappointed expression on his face as if he were saying, "In the old days it would have been standing room only, and now, look at this." I should have taken my cue from the fact that hardly any of the senior faculty showed up, but in my naïveté I chalked it up to, "Maybe he isn't as popular as he once used to be." Only many years later, by the time Teichmüller spaces made it into string theory and Bieberbach was long dead, did I look at some issues of *Deutsche Mathematik* and understand that expression on Bieberbach's face.

You may wonder what happened to Göttingen after Weyl, Courant, Noether, and others left. The man who took charge was Helmut Hasse, one of the very major number theorists of the twentieth century. In Hasse you find again that minding the store and taking care of the young people attitude we encountered in Heisenberg. Unlike Heisenberg though, Hasse had applied for party membership and was turned down because his Aryan credentials could not stand close scrutiny.

Some very disturbing stories circulate about Hasse. He reportedly wrote a letter to Élie Cartan, the great French geometer, suggesting that Cartan's son captured by the Germans would be treated well if Élie Cartan were to publicly endorse the German occupation of Paris. His letter was not deemed worthy of a reply. Then, the widow of Adrian Albert, my former dean at Chicago, told me about a paper her husband had sent to Hasse for publication in *Mathematische Annalen,* the journal that carried Hasse's, just as it never carried Emmy Noether's, name on its cover. For a while Albert didn't hear from Hasse. Then one day in the latest issue of *Mathematische Annalen* at Eckhart Library he found his paper, just as he had submitted it to Hasse, with just one minor editorial change. His own name had been removed and now two young German mathematicians were identified as its authors. Being an *American* Jew apparently didn't protect Albert from seeing himself purged from his own work, Nazi style. Plutarch, where were you when we needed you?

But what of Ernst Witt, Emmy Noether's protégé, the one who attended her lectures in full Nazi regalia? Witt established himself as a most original algebraist. After the war his ideas were widely developed, and again they showed up in string theory. As the mathematician John Thompson once explained to me, "We rigorously quote him where appropriate, but beyond that he is shunned by the community."

There is a similar case in physics, that of Pascual Jordan, whose early contributions to Quantum Mechanics and Quantum Field Theory are seminal, and who started the extremely fruitful branch of Jordan algebras in mathematics. I recall giving a colloquium at the German Electron Synchrotron Laboratory DESY in Hamburg in the late Sixties. While there, I thought I should pay a visit to Jordan, who was a professor in Hamburg and whose work I admired and occasionally built on. I asked one of my hosts how I would best get there. "Oh, that's very far from here," I was told "you're sure to get lost. You don't have time to do that." Another of the locals also discouraged me: "On a Tuesday! You won't find him there, he never comes in on Tuesdays." Then another one: "I think he is sick." It was clear that something was going on. I asked someone I knew and finally got an honest reply. "In view of his behavior under the Third Reich, we do not wish to have any contact with him." As with Bieberbach, I was being naïve. I hadn't read my Plutarch and assumed that great scientists must behave honorably.

But then, years later, my friend Feza Gürsey of Yale told me that he had met Jordan and that the man was quite weird. He was obese, afflicted by an extremely serious stutter and had no facial hair, not even eyebrows. From pictures taken in his youth this is not yet apparent, but the stutter apparently is common knowledge. So what has Jordan done? He wrote a book, *Physics in the Twentieth Century*, in which he attempts to rewrite the history of physics, portraying Einstein as a pushy Jew who tried to muscle his way into the work of certifiable Aryans but who is finally getting his comeuppance. After this book appeared, Jordan ran into Pauli, who thought very highly of him and with whom he had written some fundamental papers. Pauli took Jordan to task. "Jordan, how can you write such nonsense?" To which the cornered Jordan replied, "Pauli, how can you read such nonsense?" So, the whole thing was bogus. Why? Maybe because in a country in which they deported the Jews and the mentally disturbed you couldn't be certain they wouldn't sooner or later start rounding up other undesirables such as stutterers, or stutterers without any Aryan charm. So Jordan maybe was just covering his behind.

On the experimental physicists' side the Third Reich produced the champions of "*Deutsche Physik*," two Nobel laureates Philipp Lenard and Johannes Stark. Lenard bore grudges against many people, most importantly against J. J. Thomson, the discoverer of the electron whom he came close to accusing of plagiarism; he was also against Albert Einstein, who came up with the fundamental concept of light quantum based on the laws of the photoelectric effect, which were discovered but left unexplained by Lenard. Though a lot of great physics was built on Einstein's brilliant idea of the light quantum, with time the particular experiment that originally led Einstein to it was set aside, if not outright forgotten, by physicists. Jealousy must have raged in Lenard's heart. He hated Einstein and he hated Thomson, and, he readily generalized this into a hatred of all Jews and all English. This hatred served him well under Hitler, especially when he did not shy away from open invective against Einstein whose work he labeled "flimflam" and a "Jewish scam."

Johannes Stark shared Lenard's hatred of theoretical physics since it seemed to garner much more attention than the experimental physics the two of them practiced. There is no question, people prefer explanations to facts; only then can they understand. But many distinguished theoretical physicists were Jewish, whereas at that time few experimental physicists were. So, the experimentalists Lenard and Stark concocted *Deutsche Physik*, their own little Protocol of the Elders of Zion and decided that all theoretical physics was "Jewish physics," flimflam meant to usurp their status as German knights of the grail of physics.

CHAPTER

Spontaneous Breakdown of Human Decency

Enough of this roll call of Nazi scientists! By now the pattern is emerging fairly clearly: in times of turmoil, people abandon reason and embrace opportunism in all its many shades. But here murder was involved, mass murder. Would I, would you be ready to commit murder, or even only to condone murder, for no better reason than to see the synthetic approach to mathematics triumph, to assert the supremacy of experimental or for that matter theoretical physics? I would hope not. Would I never commit murder for any reason, under *any* circumstances? Again the answer is no. If someone threatened to kill me or my immediate family and I could eliminate this mortal threat only by killing the assailant, I would do it. I would be justified, indeed morally obligated to do it.

Were these Nazi scientists mortally threatened? Hardly. But that is not the right question. The relevant question is, did they *feel* mortally threatened? And to this question the answer, at the very least is not that obvious. Lenard, for one, was the toast of the physics community for his work on cathode rays and on the photoelectric effect. Then he got shunted aside once Einstein came up with an understanding of this effect and of its profound implications. Ironically, Einstein's breakthrough took place the very year Lenard received the Nobel Prize for his experiments on cathode rays. Everybody now toasted Einstein, and Lenard saw much of his life's work rubbed out by this Jewish upstart. This work was the very meaning of Lenard's life. Einstein had robbed him of this meaning; he had robbed him of his life, and in a certain sense killed him. This was the offense which Lenard must have felt that entitled him to exact revenge, to kill if need be.

Bieberbach was a reputable mathematician, a professor in Berlin, but in his very own Germany he ranked below Emmy Noether, below Issai Schur and other Jews as well as Jew-sympathizers like Hermann Weyl. If you have great ambitions when you start out, you see a certain life in store for you. But when this life turns out *not* to be in store for you after all, disappointment and shame become your companions in the life you *are* fated to live. You may be deeply disappointed with yourself for not living up to your own expectations, or you may get angry at others who, you believe are denying you what you are entitled to. The difference between these two perceptions is a matter of the degree to which you heed Polonius's "To thyne own self be true." In either case you are not killed, and yet the life you expected as your right is taken from you. Thinking along these lines, it becomes less incomprehensible why someone like Bieberbach would be willing to get rid of the Jewish interlopers and thereby become the leader of German mathematics. The life of a leader, of a *Führer*, is the one these foreign elements denied him, robbed from him. Oh opportunism, how wide is your reach!

Jordan may have been genuinely frightened. The Jews brought out the murderous streak in his compatriots and he, Jordan, could fall prey to it. It's all *their* fault, he may have thought, and the more quickly *they* are gotten rid of, the more quickly the murderous urge would subside, the better chance he stood of surviving.

Once the irrational takes over and becomes official policy, obedient citizens follow and literally let the *Führer* think for them, even if they are much smarter than he. Teichmüller, Witt?

It is always easy to feel that one understands the past, it all makes sense, had they but known the poor dears, it all could have been avoided. There is a much praised novel by a German judge in which an illiterate Swabian peasant woman from the Romanian Banat (incidentally that is where I come from as well; where there are Swabians, sooner or later you also find Jews) participates in a heinous war atrocity. Many years after the war she is tried, locked up, and learns to read. A little Nietzsche and a little Homer go a long way and ta-da she sees her guilt and hangs herself. Had the Germans been morally literate, is the heavy-handed moral of this metaphor, Bieberbach would have been an analytic mathematician and Lenard a Jewish Physicist.

But when you become irrational, morality is of no more use. When you are irrational, you don't just carry a chip on your shoulder, you wear the chip on your lapel as a badge of honor. The bigger the chip the better — it's the fashion of the day. You attach your chip with sturdy wire, and you feel righteous committing, or allowing others to commit, the most horrendous acts. Like you and I, most people will commit murder only under a moral imperative, and the chip provides this imperative. There are no evil men, there are only men who commit atrocities under what they are certain is a moral imperative. Without such an imperative they could not sleep at night; they'd end up hanging themselves like the Swabian woman in the judge's novel. I am sure Hitler felt justified to perpetrate the Holocaust by just such a phony moral imperative.

11

Scientists in Politics

Abdus Salam was the quintessential cosmopolitan, a globalized man, before globalization had even been thought of. He studied at Cambridge, organized the most prominent British theoretical physics group of his day at Imperial College in South Kensington and then went on to share a Nobel Prize in physics with Shelly Glashow and Steve Weinberg for their unification of the electromagnetic and weak interactions. He was the first Muslim ever to be so honored. Or was he? Strictly speaking, no, he was not a mainstream Muslim, but a member of the Ahmadi sect. His father was its leader. Yet he was very close to the Pakistani president General Ayub Khan, for whom he undertook sensitive diplomatic missions.

With the imposition of Martial Law in Pakistan in 1977, Salam's fortunes in his homeland took a turn for the worse. In the wake of the Iranian revolution, General Zia-ul-Haq, the new ruler of Pakistan, found it expedient to turn the country into an Islamic Republic. One of the first consequences of this turn of events was that, at the instigation of the mullahs, prohibitions against Ahmadis were introduced and the Ahmadis were being harassed, sometimes violently so. By then the Ahmadis had been declared heretics by the World Muslim League and they were forbidden to practice Islam and to call themselves Muslims. Despite the Nobel Prize that could have made him a role model to young Pakistanis, Abdus' contacts with the Pakistani government came to an end. He was devastated by this fanaticism. In his office at the International Center for Theoretical physics in Trieste he emotionally complained to me "Peter, can you imagine, they don't want me anymore. I went to Zia and protested and he told me 'look, Professor Salam, I know that you are as good a Muslim as I am, maybe even better, but there is nothing I can do. I promise you though, I'll see to it that nothing happens to you.' Can you imagine this? What a spineless, shortsighted man."

It hurt Salam all the more, because he had spent a great deal of time building up the ICTP in Trieste for very idealistic reasons: he hoped it would bring young scientists from the developing countries in contact with the best in world science. He knew full well that in the company of religious fanaticism, science withers. He fervently believed in the unity of civilization and kept reminding his Western friends about the time in the early thirteenth century when the tables were turned and a certain Michael the Scot came to study in the Arab world thereby beginning the transfer of knowledge that woke up a fundamentalist Europe mired in the Dark Ages. Seven centuries later, Abdus Salam founded the ICTP to start a reverse flow of knowledge from the advanced countries to the developing world. In the beginning, the institute was a great success, because Abdus was there. Its UNESCO funding was on a precarious footing and Abdus intended to raise money from those who would benefit and could afford it. This meant the Saudis and the Arabs from the other Gulf States. He was hoping for about a billion dollars in the form of an Islamic Science Foundation. He went to the Arabs and presented his case.

They politely listened to him (after all, a Nobel Prize laureate is a Nobel Prize laureate) then they gave him some medal and some insignificant amount of cash and sent him on his way.

Abdus was a good Muslim. He was careful to follow all the rules. He had two wives and he brought them both to the Nobel ceremonies. I remember being scheduled to give a talk at a conference in Trieste a few weeks after my divorce. I was such a nervous wreck that I canceled. I met Abdus in Washington a few months later. He walked over to me, patted me on the shoulder, and voiced his understanding. "A divorce can be very painful thing" he said. Then he urged me to convert to Islam. "We can have up to four wives, so if one wife doesn't work out, you still have three going. You barely notice it."

"For the time being I am satisfied with having the same number of wives as you *modulo two*, even without converting" I replied. The way I succeeded in matching Abdus' "You barely notice it" incentive was by remarrying my first wife two years after our divorce.

Feza Gürsey told me that he had been visiting Imperial College during the Sixties. He and Abdus were having a discussion, when Feza realized that he was hungry and suggested they continue over lunch. "It's Ramadan, you are Muslim, you are not supposed to eat till sundown" Abdus reproached him. "When I work, I work and for that I need to feed myself, so if you don't want to come, I'll go alone." And Feza went to the Imperial College faculty club where *entre nous,* they served food which is to English food, what English food is to French food. When he returned, he went right to Abdus' office to continue the discussion. He knocked on the door more as a formality, Feza was a gentleman through and through, and then he opened the door without further ado. That moment a drawer slammed shut in Abdus' desk. Feza proceeded to ask Abdus a question that had preoccupied him over lunch. Abdus sheepishly looked at Feza, but couldn't answer, his mouth was full. He swallowed and then opened the drawer "It's only a banana. A banana! See?" Feza understood. He was a Muslim from secular Turkey and had nothing to prove, where religion was concerned. Abdus was an Ahmadi from Pakistan, a very big difference indeed.

Ironically, the most famous of Abdus Salam's students is Yuval Ne'eman, a native of what then became Israel, destined to become minister of science in Menahem Begin's Likud cabinet. Yuval had a very high position in Israeli intelligence. In the Fifties he was dispatched as defense attaché to the London embassy to negotiate the purchase of two submarines. The negotiations were routine and rather slow, giving Yuval ample time to restart his abandoned studies in theoretical physics. The University of London itself has a number of colleges which offer physics and then there was Imperial College, which happened to be the nearest to the Israeli Embassy. So Imperial College it was. Under Salam's guidance, Yuval completed a thesis which introduced a picture of strongly interacting particles, or hadrons, called "the eightfold way" by Murray Gell-Mann, who independently discovered it. The success of this eightfold way led directly to the introduction of quarks by Gell-Mann and George Zweig a few years later. Yuval was now a fulltime theoretical physicist.

During the Six Day War Yuval wanted to be of help, but having been away for so long, he was put in charge of press releases. He told me that this job was rather funny. On the one hand, the Arab media were claiming victory upon major victory and heralding the imminent destruction of the State of Israel. This in turn drove the Jews in the Diaspora to make large and badly needed financial contributions to the State of Israel. On the other hand, with the destruction of the whole Egyptian Air Force on the first day of fighting, the war had already been won by Israel. So it became Yuval's task not to contradict the Arab claims and draw the fighting out as long as possible, both to keep the money flowing and to clearly crystallize the military situation. He managed to do this for only six days, as the name of the war shows.

Yuval and I often crossed paths scientifically, and we became friends. Besides physics we often got into politics. Yuval was a hawk. He strongly opposed the return of the Sinai to Egypt, and he resigned his high official position in protest. He would complain about Yitzhak Rabin and on military matters clearly sided with Arik Sharon. One winter afternoon we walked back from City College to Midtown Manhattan, crossing some rather crime-ridden neighborhoods in search of a cab. Yuval was holding forth about all the mistakes that were being made and about what he considered to be the right policies for the State of Israel. During the ensuing cab ride, just as we were passing the Museum of Natural History on the West side of Central Park, I asked him why he doesn't go into politics. He sounded surprised at this question, although I got the distinct impression that the thought had already crossed his mind and that he was using me as a sounding board. In any case, soon thereafter Yuval founded the very right wing Tehiya party, he got elected to the Knesset.

He was much reviled in the physics community for his politics, which, to my mind, was consistent and rational, once you granted its premises. Yuval's main issue was what should be done with the territories Israel acquired during the Six Day War. If a country is attacked, and repels the attack it can follow three paths. Obviously it can start peace negotiations with the defeated aggressors and use the land as a bargaining chip. It can outright annex the land, evict its inhabitants, colonize the land and hope for the best. The third possibility is to occupy the land and act as an occupying power for an extended period of time and try to colonize the land piecemeal.

In my opinion, this third possibility is the one any sensible government will avoid, for it breeds resentment and in the end inevitable rebellion. When the Six Day War ended in a decisive Israeli victory, I had hoped that this would lead to negotiations. Apparently, some negotiations were attempted, but led nowhere. Yuval, by contrast, clearly favored annexation. Whether one does or does not agree with him on this choice, the policies he advocated were all its clear, indeed rigorous consequences. It was all very rational.

In his years as minister of Science he proposed some very imaginative projects, chief among them was a canal linking the Mediterranean with the Dead Sea. But ultimately the hand he had been dealt involved occupation, and we have seen where this has led.

Yuval was by no means the first scientist statesman. René Painlevé, the famous French mathematician had served as his country's Prime Minister, and Sir Isaac Newton had been appointed Warden and later Master of the Mint and made a lasting contribution there by introducing the serration of coins. I guess his numismatic fame is comparable to his fame in physics and mathematics, except that very few people, though admittedly rich people, care about numismatics.

Einstein too was almost dragged into active politics. When the State of Israel was founded in 1948, he was approached to become its first president. Leading politicians from Israel came to Einstein's Princeton home to make the offer. Einstein wisely declined. According to General Yigal Yadin, though, the Israelis had a contingency plan, in case Einstein *did* accept their offer.

But Einstein had a keen interest in and a solid understanding of politics. One of the first stories about a physicist that I have ever heard has Einstein as the students' guest of honor at a University of Berlin evening discussion in the economically painful and politically ominous Twenties. First a student gave an impassioned speech about how economics determines history, to the exclusion of any meaningful role for the individual. When he was done, the speaker turned to the great man for his expected approval, but Einstein completely disagreed with what the student had just said and to make his point, he addressed a young student sitting in the middle of the hall. The student rose. He was a rather gaunt lad, obviously undernourished, wearing a threadbare coat, and not cleanly shaven, in fact rather slovenly. The large straight scar across his right cheek marked him as a member of one of those right-wing dueling fraternities. Einstein asked him whether he had enough money to support himself. Not surprisingly, the young man answered that he could barely subsist.

"Wouldn't a monthly stipend of" here Einstein named a sum — "go a long way towards solving your problems?"

"Sure it would, but where in this day and age am I to get such a stipend?"

"As it happens, I have been asked by the Socialist Students' Association to nominate someone for precisely such a stipend. Give me your name and I will nominate you."

"Never! I would rather starve than accept money from the socialists," said the irate student, and he sat down.

"So much for history being determined by economics," Einstein concluded with a smile.

Einstein died half a century ago, Yuval Ne'eman's Tehiya party has lost all its seats in the Knesset and he died last year. Abdus Salam died in 1996 following a long battle with Parkinson's disease.

12

Jews in Science. The Backlash

There is a joke often told by physicists: communication is established between humans here on earth and extraterrestrials living in a galaxy made of antimatter. It is found that in that anti-world they have anti-science, anti-mathematics, and anti-physics. Earthbound physicists get a description of an anti-physics anti-laboratory, and lo and behold, they find it is filled with anti-Semites.

Jews are represented in present day science in much larger numbers than one might expect, based on their share of the world's population. It was not always so. At its birth, modern science was an exclusively non-Jewish endeavor. Copernicus, Kepler, Galileo, Newton, none of them were Jews. Only after the emancipation of the Jews, largely as a result of the Enlightenment and of the French Revolution, do significant Jewish scientists and artists appear on the scene. The first to show up are mathematicians: Jacobi, Sylvester, Eisenstein, Kronecker, ... composers and poets: Mendelssohn-Bartholdy, Meyerbeer, Halévy, Offenbach, Mahler, Heine.... These arts and sciences require very little in way of apparatus and facilities: no lab, no studio.

From these beginnings, over the nineteenth and twentieth centuries, Jews branched out into all scientific and artistic fields, largely because in these activities the output is something entirely unique, its value clearly recognizable to all practitioners and to the public at large and therefore unaffected by the ethnicity of its creator. No mathematician can do without Georg Cantor's set theory, whether Cantor, a practicing Lutheran, had Jewish ancestry — as claimed even by himself — or not. Nor can a musician do without *Songs of a Wayfarer.* It's not like you have a choice between Jewish set theory and the Aryan kind, or between the songs of a Jewish and those of an Aryan wayfarer. But the rise of Jews in the arts and sciences did not occur smoothly without obstacles.

In the beginning the appearance of Jews in the arts and in the sciences was allowed to pass as some kind of transient phenomenon. So say, a Jew takes a crack at an equation or writes a song — so what? But before long the backlash did not fail to arrive. No lesser a musician than Richard Wagner spent a good deal of time ranting and raving against Jewish composers, not only through vitriolic pamphlets, but even through his music in which most villains, Alberich, Mime, Beckmesser, Klingsor, all have Jewish musical mannerisms. Ironically, a good case can be made that these villains are Wagner's most compelling creations, chalk one up for racism. The first act of *Siegfried*, very likely the best thing Wagner ever composed (I was much pleased to discover that I share this view with Dmitrii Shostakovich the celebrated Russian composer) is a blatant attempt at mocking the Jew Mime. In the end this makes his character much more interesting than all the virtuous Aryans Wagner claims his music is really about.

The backlash in physics and in mathematics — at least outside the Third Reich — took a more "civilized", but hardly much less harmful form.

For instance, George Birkhoff the first world-famous American mathematician was, in the words of Albert Einstein,

"one of the world's great anti-Semites" (no pun intended?).
This was the academic counterpart of Henry Ford's and
John D. Rockefeller's well publicized anti-Semitism. Did it
have any serious consequences? Yes and no.

Take the case of Murray Gell-Mann, certainly one of
the most brilliant minds of the second half of the twentieth
century. He did his undergraduate studies at Yale; for
graduate studies he went on to M.I.T., having been rejected
by Princeton. Why would Princeton reject an applicant of
his caliber? Arthur Wightman, the well-known Princeton
axiomatic quantum field theorist searched the record and
found that it wasn't entirely Princeton's fault. Gell-Mann's
Yale recommendation letter portrayed him as a pushy fellow, in other words a pushy Jew, who wasn't as good as he
thought he was. That Murray was not a wallflower is hardly
news, but given his achievements, it is hard to see how he
could possibly have overestimated his worth.

One is reminded of the old story about Henry
Augustus Rowland, the first internationally acclaimed
American experimental physicist after Benjamin Franklin.
In the late nineteenth century he was professor of physics
at Johns Hopkins University and one day he was called to
give expert testimony in a court of law. The questioning
attorney asked him, "Professor Rowland, who in your opinion is the best physicist in the United States?" To which he
replied "I am, sir." The next day his enraged colleagues took
Rowland to task about his lack of modesty. "But I was
under oath," a sincere but defensive Rowland replied.

Still, prewar American anti-Semitism like the prewar
British variety on which it modeled itself is what one may
call, somewhat oxymoronically, civilized or gentlemanly
anti-Semitism — no pogroms, no gas chambers. The most
painful consequences of officially condoned anti-Semitism
in the sciences, outside of Germany, were felt in the Soviet
Union. In mathematics this is clearly associated with the

names of the topologist Lev Semenovich Pontryagin, one of the major mathematical geniuses to emerge from twentieth century Russia, and of the brilliant analytic number theorist Ivan Matveyevich Vinogradov. Pontryagin's anti-Semitism is all the more fascinating to explore because it is amply documented and because it very clearly exhibits the irrational aspect of bigotry.

A good place to start is Pontryagin's scientific auto-biography. When a well-known Soviet mathematician reached the age of seventy, custom dictated that one of his closest collaborators or pupils would publish a scientific biography in the journal *Uspekhi Matematicheskikh Nauk* (Progress in Mathematical Sciences). In 1978 Pontryagin turned seventy, and *he* himself went on to write his own biography for *Uspekhi*. The copies of this *Uspekhi* issue destined for American subscribers arrived mutilated. The autobiography, announced in the contents of the issue was missing — the pages had been literally cut out by the Soviet censor for political reasons. The autobiography was found so unabashedly anti-Semitic that it was feared it might discredit the image the Soviet Union was trying so hard to project abroad.

Nevertheless, copies of the Pontryagin autobiography were smuggled out of the Soviet Union. The decisive moment in Pontryagin's life was a childhood accident that left him blind at the age of 13. While blindness did not dull his deep mathematical intuition, it understandably heightened his suspiciousness, an attitude already plentiful in any Soviet citizen, as a simple matter of survival. He studied in Moscow with the topologist P.S. Aleksandrov. A.N. Kolmogorov, another dominant figure of mid-century Russian mathematics, was his rival, and unquestionably his peer.

As Pontryagin tells it, Kolmogorov hovered at all crucial moments over Pontryagin's work, belittling it and trying to prove it wrong. As if this wasn't hard enough to bear, Pontryagin was riddled with fear of nearly everyone. In the Brezhnev era, in which the autobiography appeared, Pontryagin willingly admits that he first learned of the War from Molotov's speech and was sure the Soviet Union would lose it. Such a defeatist attitude coupled with the mention of a non-person like Molotov, could easily send a lesser Soviet citizen to Siberia, even as late as the Brezhnev Period of Stagnation, with Stalin resting peacefully in the Kremlin wall. Yet he doesn't hesitate to admit to all this. There is a certain spark and daring to the old fellow.

The fear of war led him into a disastrous marriage to a woman who is left unnamed and who apparently was Jewish. After the War they got divorced and Pontryagin was reborn as a card-carrying anti-Semite. His mentor in this new vocation was Vinogradov. Together they slowly took over the Soviet mathematical establishment. So what was Pontryagin's most important task? To rid the main Soviet publishing house *Nauka* of the "monotonous repetition of the authors' names," in other words of the large number of authors whose names sounded Jewish. Careful mathematician that he was, he implemented an equivalent policy with the award of postgraduate degrees. When sitting on the doctoral exam committee of a very talented mathematician named Vinberg, he rejected the candidate and then boasted "I flunked the Jew," only to be told afterwards that Vinberg wasn't Jewish at all, but of good Northern stock courtesy of Peter the Great. He made amends and Northern Vinberg got his Kandidat Matematicheskikh Nauk title, roughly the equivalent of a tenured associate professorship in the US.

If only all stories had as happy an *in situ* ending as Vinberg's. But Vinberg's own student, the great algebraist Viktor Kac, had to leave the Soviet Union because of

Pontryagin. I met Kac a few days after he arrived in the U.S. in 1976, when we were both invited to talk at a conference at the University of Virginia, I asked him whether it had been easy for him to take the step of leaving the Soviet Union. There was no question that living in the States was much, much better than living in the Soviet Union. On the other hand, one's scientific creativity depends on many intangibles: on the relations with one's colleagues, on the meaning of the smile on a colleague's face just after he has been told of a new idea, on one's familiarity with the work environment, on one's knowledge of the location of a certain book in the library, on one's knowledge of the exact path to the library so that while on the way, one can let one's thoughts continue undisturbed on their own path. Going to another place, admittedly a better, much more pleasant place, does not mean that it will all work out for the best. It is a very big risk, and yet Kac, a man who in the rugged Soviet terrain had produced superb work — he had already started whole branches of algebra — was taking this risk. Why?

"In Moscow I had to spend a lot of time just getting my papers published. I would submit a paper and six months later, Pontryagin would reject it without giving a reason beyond the meaningless statement that the paper was not 'up to the standards of Soviet mathematics.' By this he meant no more than that my name sounded Jewish."

At least this time around, the old fellow had it right, for, unlike Vinberg, Kac is *not* of "good Northern stock." Incidentally, Kac has continued doing splendid work at M.I.T. He obviously has found his way to the Cambridge library and has learned how to read a Yankee smile.

In his autobiography Pontryagin sees Jewish conspiracies all over the place, even in the International Mathematical Union. When serving on the IMU's Executive Committee he claims that "An attempt was made by the

Zionists to take the IMU in their hands. They tried to raise Professor N. Jacobson, a mediocre scientist, but an aggressive Zionist, to the Presidency of the IMU. I managed to repel the attack." In classical Soviet style he refers to Jacobson as a Zionist where he clearly means Jew, since he most certainly was not privy to Jacobson's politics. As to the insulting evaluation he makes of Jacobson's contributions, Jacobson was a leading algebraist at Yale and the author of some famous papers and books. Was he in the same league as Pontryagin? No, but then how many mathematicians were? Now think of this ten-member Executive Committee in session, everyone speaking English, a language Pontryagin spoke not all that well, and followed probably worse when spoken idiomatically at conversational speed. Think of this blind man listening to all the incomprehensible, and as such boring, babble coming from nine men. Who was to say he was not at a session of the Elders of Zion?

So, for whatever reasons, Pontryagin was a virulent anti-Semite. But why did no one in the Soviet Union try to put an end to his abuses? For instance, another mathematician who was his peer, say Kolmogorov, as great a mathematical genius as Pontryagin himself, or Pontryagin's teacher P.S. Aleksandrov. The very dramatic answer to this very natural question is that they could not, for Pavel Sergeyevich Aleksandrov and Andrei Nikolayevich Kolmogorov were openly living together. Everybody knew it, and it was tolerated — anyway tolerated within certain bounds. Had they taken on Pontryagin, they may have been writing their tickets to Siberia. And then they say that mathematics is such an abstract, remote business.

Now Pontryagin is dead, so are Kolmogorov and Aleksandrov, and more importantly, so is the Soviet Union. The Elders of Zion have finally been put out of circulation, right? Let me restrict the statement somewhat. The Elders of Zion have finally been put out of circulation at least in the Russian mathematical community, right? Wrong! The latest embodiment of that old forgery now parades as *Russophobia*. It is the basic tract of the new Russian anti-Semites, written by Igor Rostislavovich Shafarevich, a major Russian mathematical genius.

Stalin and the Quantum

Not only did fascism and communism, the twin evils of the twentieth century, agree to a large degree on how to handle the Jews, but the two ideologies also had similar attitudes towards quantum theory. The Nazis hated it because they saw it as Jewish physics, the Soviets hated it because they perceived it as being at odds with dialectic materialism, the ideological basis of the dictatorship of the proletariat.

The feature Stalin and his commissars objected to, was the probabilistic interpretation of quantum mechanics developed by Heisenberg and Bohr in Copenhagen in the 1920s. According to this Copenhagen interpretation, for a series of identical experiments one can predict the probability of any particular outcome, but not which outcome will be found in each individual experiment. For example, if one shoots a beam of electrons at a target made of protons, then one can predict the probability that the incoming electrons will be deflected by any given angle (or more precisely into any given angular region) but not whether a given electron, say the twenty-seventh electron to hit the target, will or will not be deflected by that angle. It's very much like with a fair coin toss: one can predict the probability of tails as outcome to be 50%, but not whether on the twenty-seventh toss the outcome will be tails.

Einstein himself objected to this feature of the Copenhagen interpretation. "God does not play dice" was how he famously put it. Stalin could not have cared less whether God plays dice or not — what bothered him was the appearance that the electrons had a "free will." He preferred his electrons obedient. Moreover, even communism, which in the Stalinist credo was the one hundred percent certain outcome of social evolution, could be undone if electrons and protons and atoms and systems of atoms such as Soviet citizens, could retain even a trace of free will. So, it was decreed that the Copenhagen interpretation contradicted "scientific materialism" and therefore was idealistic, which is stalinese for *verboten*. Trouble was, strictly speaking, this made A-bombs, the functioning of which can only be understood in this "idealistic" fashion, *verboten* as well, much as if a kind of ideological arms control had been unwittingly activated.

Soviet philosophers had a field day. They found heresy upon dangerous heresy lurking in Lev Landau and E.M. Lifshitz's unsurpassed textbook on quantum mechanics, and the possibility had to be contemplated by these two physicists that they might be looking at a stay in Siberia.

Something had to be done. Something *was* done, and the way the Copenhagen interpretation was reconciled with Marx very much resembles the way the Iranian ayatollahs reconciled prostitution with the Koran. The ayatollahs ruled that the Koran specifically allows temporary marriages and that therefore a male visitor to a "house of chastity," is not there to avail himself of sexual services for a fee. No, he is there to enter into a temporary marriage. He marries a chaste woman for anywhere between ten minutes and an hour, depending on his assets and his endowment, and when the negotiated term expires he leaves as an unmarried man, after having not paid a fee, but given the woman a present. With ten or more such temporary husbands a day, this chaste woman can live a truly wholesome life.

Marx was taken to the Copenhagen altar in much the same spirit. The only surprise was that the ayatollah officiating at Marx's house of chastity, was none other than V.A. Fock, one of the greatest Russian theoretical physicists of the twentieth century.

In the late Fifties Fock wrote an article for a Soviet booklet, dedicated to the philosophical problems of modern physics in which he reported about a conference during which he had had the occasion to catch both Heisenberg and Bohr, the original proponents of the Copenhagen interpretation. He had asked them point-blank whether their interpretation contradicted dialectic materialism, and they assured him that to the best of their knowledge there was no contradiction. If even its proponents openly admitted that the Copenhagen interpretation does not contradict the tenets of Marxism-Leninism, the official Soviet ideology, then, Fock went on, there was no need to discard this essential part of modern physics as idealistic, and in one fell swoop he felt he had rehabilitated quantum mechanics for Soviet use. Clever!

I remember reading Fock's paper while still a student in Romania, and I was awed by this man's chutzpah. To understand this reaction, keep in mind that in the course on Quantum Mechanics I had attended, the lecture on the uncertainty relations (according to which the more accurately one determines a particle's position the more uncertain one becomes about its momentum — or if you wish, about its velocity — and vice versa), was prefaced, admittedly with a twinkle in the professor's eyes, with the admonition that these were unacceptable "idealistic" misinterpretations produced by Heisenberg "at the order of his

American imperialist masters." I remember thinking I wish they'd order me to produce something like that. Now here came Fock with the idea that Stalin could have ordered these same relations, had he been as smart as Heisenberg's American imperialist masters, Marx and Lenin would not have objected.

In 1979, at the Einstein Centennial Conference in Jerusalem, Loren Graham, the well-known M.I.T. historian of science, gave a talk about Fock who was by then deceased. The talk was very interesting and very well researched, but there was no mention of this remarkable paper.

During a reception I asked Graham whether he was aware of the paper, and if so, why he didn't mention it. He in turn asked me to first tell him what I knew about it. I told him more or less what I have written here. He smiled, looked at me, and asked "Am I correct to infer that you did not know Fock?" I answered that he was more or less correct, for I had met Fock once in Trieste in 1968, but we exchanged only a few words. What I remembered best had been his elaborate Russian-made hearing aid, which he kept tuning according to his interest in the conversation going on around him. Fock had been a ballistics expert assigned to an artillery unit in World War I and was for all practical purposes deaf.

At this Loren Graham told me that he had heard many reports on that famous paper, but that the reports changed depending on the closeness to Fock of the one doing the reporting. People like me who barely knew Fock indeed found his paper daring — though Stalin himself was no longer around by the time Fock wrote it, Khrushchev, his successor, was hardly more tolerant when it came to ideological matters — and saw Fock as a hero who had taken a great risk in the interest of science. But if you talked to people close to Fock, like his assistants, you get an entirely different story. According to them, Fock was a true believer in Marxism-Leninism and a great physicist at the same time. Physics and Marxism-Leninism were the two great loves of his life. That these two loves contradicted each other upset him no end. When he went to Copenhagen, he decided to use the occasion to resolve this unhappy state of affairs. He did approach Bohr and Heisenberg with the goal of convincing them of the validity and ultimate inescapability of Marxism-Leninism — in short, he was doing missionary work on behalf of his religion. He was sure he had been successful, and he was prouder of this achievement than of virtually any of his great physics discoveries. Thus with the two Copenhagen giants freshly converted true-believers in the religion of the hirsute bearded fellow and of the bald goateed one, there was finally peace between the two main loves of Fock's life, and he could go about working on quantum mechanics without even a tinge of guilt.

Heisenberg in his memoirs mentions this meeting with Fock as a kind of nuisance interfering with an attempt at a postwar Bohr-Heisenberg fence mending. It is all so ironic, if you think of it. Heisenberg tries to make peace with Bohr, and here comes this Russian with his Marx and with his Lenin and interferes. Yet, there is Fock honestly believing to have reconciled the Copenhagen interpretation with Marxism-Leninism, while all he did was to stumble upon the right formula to turn Soviet universities that teach quantum mechanics into houses of chastity in full compliance with the teachings of the holiest Red prophets.

Mister Sakharov
Goes to Kiev

In 1970, the major biannual conference in High Energy physics, the so-called Rochester Conference, took place in Kiev. I was a member of the U.S. delegation. Shortly after I checked in at the Ukraina Hotel in Kiev, Academician Yakov Borisovich Zeldovich invited me to his suite. He wanted to find out about some work I had done. We sat down at the table in the salon of his suite and I started telling him my results and writing formulae on a piece of paper with his Russian ballpoint pen. He would then take the pen from my hand and ask questions, writing out his own formulae. After a while we switched subjects, and he started telling me about some work of his regarding what he called "induced electrodynamics," his variation on a theme by his friend Andrei Sakharov.

I wanted to ask my own questions, but this passionate and intense man was so carried away with his own argument that he tolerated no interruption, and he held the floor by simply denying me access to the one pen we were both using. At this point I got out my maroon fountain pen, a Sheaffer Lifetime that I had purchased at the

University of Chicago bookstore for thirty dollars. Suddenly Zeldovich stopped mid-argument, very carefully took the pen from my hand, and in awe exclaimed

"Par-kerre!"

"No, Sheaffer" I replied.

"She-fer" he echoed with even greater awe.

With deep respect for an other man's fountain pen, he asked me whether he could try it out. From where I come from, you lend your wife to another man before you allow him to use your fountain pen, yet his excitement was such that I couldn't bring myself to turn down his request. So we finished the discussion of his variation on Sakharov's theme and of its attendant difficulties with him using my "She-fer" and me using his Russian ballpoint. With my fountain pen thus adulterated, I thought I might as well give it to him altogether as a gift. So I told him "Look Zeldovich, you seem to enjoy this pen very much. Allow me to offer it to you as a gift from someone who also loves fountain pens. Incidentally, in my room I have a supply of cartridges for it which I can give you later."

At this, Zeldovich got visibly upset, jumped up and walked to the wall. Facing the wall, his back to me, he started in a very slow, loud, formal and heavily accented English, "Thank you very much, Professor Freund, for offering me this beautiful gift, which I cannot accept. I am giving back to you the *She-fer* fountain pen." Then he turned around and handed me the pen; it was clear to me and it was clear to him that it was clear to me that he had just addressed a bug in the wall and cleared himself in advance of any conceivable charges of having accepted gifts from Americans. He then invited me to the small balcony, and there, facing the street, told me softly that Sakharov was expected in Kiev and would be giving an unscheduled talk about induced gravity. He'd let me know later precisely when and where.

If all this sounds surreal, keep in mind that we were in Brezhnev's USSR and that, unbeknownst to me at that time, Zeldovich, along with Sakharov, had co-fathered the Soviet H-bomb. Otherwise, Zeldovich was an extremely broad physicist. He had made an important contribution, later to be rediscovered by Feynman and Gell–Mann, to the theory of weak interactions. He was also one of the seminal thinkers in modern cosmology, where to this day his ideas hold center stage.

This breadth should not come as a great surprise, for he was a student of Lev Davidovich Landau, the greatest Russian physicist ever and a very broad scientist. In fact, the Landau school is one of the greatest marvels of Soviet science.

Landau was well known for his acerbic wit. Some time after the 1962 car accident, which badly impaired his short-term memory and put an end to his brilliant scientific career, a collaborator came to visit him and asked the polite question,

"How are you feeling this morning?"

"Not as good as I used to feel, but still better than Zeldovich" was Landau's reply. I had heard about this quip before I met Zeldovich in Kiev, but found him extremely impressive all the same.

The time for the Sakharov talk came. It was a very weird occasion. The talk had not been officially scheduled. Though not yet in internal exile, Sakharov was already *persona non grata* on account of his paper on the convergence of the communist and capitalist systems, widely circulated in the West. He was arguing that under cultural and technological pressure the communist and capitalist systems will become ever more alike in order to survive, until they end up essentially identical and therefore at peace with each other. This was a far cry from Stalin's inevitability of communism and Khrushchev's "We shall bury you!", both still part of Soviet dogma.

In Kiev, Sakharov came as a scientist and his talk was to be purely scientific, no convergence. When I reached the area where the talk would take place, most seats were already taken by men without conference badges, dressed alike and of very similar appearance, all with the same type of haircut. You couldn't even enter the building without a badge, so these were clearly not physicists. You didn't have to be a theoretical physicist — even a rocket scientist could have deduced that these men were all KGB. There were very few people from abroad, probably all invited by Zeldovich. Abdus Salam was there, and so was Valentine Telegdi, Maurice Goldhaber and a few others.

Sakharov wanted to present his ideas to the foreigners, no doubt about that. It would therefore have made a lot of sense to have him talk in English. I don't know how good his command of English was, but he decided to speak in Russian with Zeldovich translating into English. I think this was for the benefit of the KGB contingent, so they could clearly jot down the words he spoke without fantasizing all sorts of treasonous intent in what he said. The trouble was that a passionate, excitable man like Zeldovich was not well suited for the translator's job. Sakharov's first two sentences were carefully rendered in English. Then Sakharov spoke for two minutes and it took Zeldovich four minutes to translate. Then the discrepancy between the two lectures, Sakharov's original and Zeldovich's translation, kept growing.

Sakharov was speaking about gravity (I understand Russian) while his translator, "for simplicity's sake," was speaking about electromagnetism. Zeldovich was really giving his own talk about his own work, the work he had related to me in his suite. Suddenly Zeldovich started writing formulae on the blackboard while Sakharov was not writing anything. An ever wider smile was settling on Sakharov's face, and at one point he interrupted the translation with a remark delivered in broken, if grammatically correct English — that the beautiful translation provided by his friend was straying very far from what he had wanted to say. Zeldovich was so involved in his own lecture that he paid scant attention to this interruption and just went on. It was a marvelous spectacle, even if the amusing ideas under discussion were somewhat flawed — a crucial sign was coming out wrong and not on account of the translation.

After the talk, we all got to meet Sakharov. He was an aristocratic figure out of Tolstoy, an officer in the Imperial army, but dressed in crumpled clothes. We exchanged meaningful glances, but nothing politically incorrect was said. The young men in the audience were all there to see to it that this would not happen.

15

Mr. Yang also Goes to Kiev

On the road to the Kiev conference, my wife and I had stopped in Moscow, and during the intermission of a mediocre Bolshoi Theatre performance of Rimsky-Korsakov's mediocre *The Tsar's Bride*, we ran into Frank Yang in the lobby. We got together — Frank, my wife, and I, and a young man with bushy eyebrows, obviously a Russian. We found it rather awkward that Frank did not bother to introduce him to us. The young man was eagerly listening to our conversation and then, some ten minutes into it, suddenly left without saying a word. Frank asked us why we hadn't introduced the young man to him. We told him that we didn't know the man and had assumed all along that he was with Frank.

At this point we all finally got it. An obviously American couple — I had a lot more hair on my head in those days and it was long to boot — and an obviously Chinese man meet at the opera; it is high time for the organs of state security to step in, who knows what these three may be up to? Then, when the vigilant agent assessed the situation to be harmless, he simply left. How's that for the friendly dictatorship of the proletariat? Anyway, we decided to continue our sightseeing together. I remember walking to Red Square for the first time, with Saint Basil's, that colorful, energetic, defiant building coming into view and Frank, an accomplished photographer, preparing his camera to snap a picture while muttering, "How vulgar, how vulgar." I think it was his gut reaction to Saint Basil's.

Frank had a lot of trouble in Moscow. He was literally being harassed by the authorities; they misplaced his passport, and all kinds of other complications came his way. He had gone to Moscow State University to visit Igor Tamm, the Russian Nobel Prize winner and Sakharov's mentor. Frank entered the lobby and a cleaning woman came at him, chasing him out of the building with her broom. The Russians had broken with China some years earlier, but by 1970 a complete anti-Chinese transition had taken place, and hatred of the Chinese was openly encouraged in the Soviet Union. In the mistaken belief that China-hatred is something universal, the organizers of the Kiev conference started the banquet by assembling all attendants in a movie theatre in which they screened a long and blatantly anti-Chinese animated picture. In it vicious creatures with long black braids and slanted eyes were invading the land of peaceful and patient Russians, only to be thrown out when these good guys finally decided that they had had enough. I remember seeing an uncharacteristically angry and outraged Frank Yang storming out from the banquet.

Frank Yang's role in theoretical physics cannot be overstated. He discovered the kind of field theory which provides the mold that fits the strong and electroweak interactions. These theories are known as nonabelian gauge theories, or Yang-Mills theories. His discovery with T.D. Lee of the handedness of the laws of nature caused a major revolution in our understanding of these laws, and they shared a Nobel Prize for it. This led to a number of further major discoveries by Lee, Yang, Oehme, Salam, and Landau. Yang also made some very important contributions to statistical physics with Lee and to some beautiful topics in mathematical physics.

Yang's collaboration with T.D. Lee is particularly remarkable. They had met already in China, then reconnected as graduate students at the University of Chicago and kept in touch after that. They obviously complemented each other marvelously and they looked inseparable. Then something happened. It is generally believed that the straw that broke the back of this extremely successful collaboration was a 1962 *New Yorker* profile of Lee and Yang by Jeremy Bernstein. The problem this article is claimed to have exacerbated is the mismatch between the alphabetic ordering of the names Lee and Yang and the ordering of these two physicists by age seniority, which is important in Chinese culture. Even if this mismatch did matter in the breakup, it would be utterly naïve to believe that a mere reversal of the alphabetic ordering, say by spelling Frank's last name with an I instead of the Y could possibly have solved the problem. Scientific collaborations are akin to marriages, or temporary marriages. Their breakup is not unlike a divorce and rarely avoids acrimony.

What motivates a collaboration in the first place? Of course there are experiments which with the best of intentions cannot be carried out by a single scientist. Here collaborations are of essence. Think of the large-scale high energy physics experiments budgeted in the hundreds of millions of dollars in which hundreds of physicists from scores of institutions participate. Some of the collaborators don't even get to meet in person. I once heard a talk by a member of such a large collaboration who in his first slide indicated that "553 plus or minus 1 physicists" had carried out the experiment. He then explained that two of the original collaborators had not been heard from in a very long time, and no one knew whether either of them was still involved. In the poet's words, "*Sie sind der Welt abhanden gekommen*" [the world has lost track of them].

That is not the kind of collaboration I have in mind. Rather, I am thinking of the simplest case of two scientists working together. What motivates these two? Could either of them have pulled the proposed research off by himself? Very likely so. What then gets them to do it together? Let me pursue the analogy with marriage. A couple is enthusiastic for the beauty of life. They get together to procreate and throw their genes in the universal pool, thus propagating life on earth. Of course, neither could have done this alone, except possibly by cloning. Yet either of them could have traveled to some dire corner of the earth and taken care of needy children there, thereby rendering an immense service to humanity. But, egotistically, they opt for creating their *own* children instead. Even so, they both get their irreplaceable roles acknowledged, the child has a mother and of necessity it also has a father. Barring any hanky-panky on the wife's part, the husband holds an equal share in these children.

Now take two physicists — call them A and B. They are both enthusiastic for the beauty of physics and want to see this science develop. A and B get together, pool their resources as it were, and create some new work. Their papers are their babies. Do they have an equal share in these papers? The alphabetic (or lexicographic, if needed) listing of their names implies that they do. But in his heart of hearts B knows that it was he who wrote on the blackboard that last equation, the one that is the crux of the whole matter. Sure, even A remembers that. But just before B wrote that equation, A had made a remark that drew attention to an idea they had discussed a week earlier, and it was a variant of this very idea that led B to go to the blackboard and write down that equation. Scientific discovery in a collaborative effort is replete with such intangibles. Once one enters the collaboration, one implicitly concedes equal partner rights to the other.

Yet over time such "heart of hearts" insights may accumulate and build up rancor, which upon a minor perturbation, the one straw, may bring the collaboration down. It all depends on the personalities of the collaborators, on their views of themselves and of each other, on their sense of fairness, on the way they are regarded by the community and on who knows what else. Yes, along with the deep love for the science, a large dose of ego, just as in the case of marriage, is also involved. Anyone claiming to be sure that it is all just an idealistic love of beauty is but revealing his own naiveté, I would say.

The personalities of Lee and Yang, or Yang and Lee, even as they appear to a third party like me, could not be more different. Yang is a refined aristocratic man, very precise, not prone to emotional outbursts, reserved, someone who communicates through the subtlety of a smile. Lee is an earthy character who enjoys a healthy bout of laughter, who loves to crack a joke, bursting with energy, driven. They worked together for the better part of a decade, and it is not inconceivable that in the end for each of them his heart of hearts was full of assorted perceived injuries. Maybe what awaits an explanation is not why this magnificent collaboration ended up in shambles, but rather why two personalities so different hooked up with each other in the first place. Be that as it may, we are the better off for their having done so.

I got the most realistic account of how this breakup occurred from the great French-American mathematician André Weil. He phrased it as a morality play. At one point in the late Fifties John Milnor, who had received the Fields medal, the mathematicians' equivalent of a Nobel Prize for his discovery of exotic spheres and who was a professor at Princeton University, indicated to Weil that he would like to move to the Institute for Advanced Study, also in Princeton. Weil immediately started the ball rolling at the Institute. J. Robert Oppenheimer, the director of the Institute, was outraged on account of a gentleman's agreement between the two Princeton institutions not to raid each other's faculty. Weil, who did not have much love left for Oppie — in the scientific community Oppenheimer went by this nickname — in whom he saw a crass politician intent on a dictatorship of the physicists, went on all the same with the Milnor appointment. Oppie did not want the matter to even come up at a faculty meeting with all the possible embarrassing consequences, and he decided to send Yang and Lee, both then at the Institute, to convince Weil to drop the matter. They showed up in Weil's office, with Yang doing the talking and Lee seconding him. They agreed that Milnor did want to come, but they insisted that the peace of the whole Princeton science community was at stake. With Weil rather stubborn and contrarian, Yang finally turned philosophical.

"You know, I come from an old civilization where the wisdom of millennia has led to the development of subtle forms of conflict avoidance. Believe me that the proper thing for you to do is to drop the matter."

"Unlike you," Weil, the Frenchman, replied "I come from a younger civilization and I am persuaded that the right thing to do is to pursue this matter to its conclusion." The proposal did come up in a faculty meeting, and while it did not pass in the first round for precisely the kind of reasons raised by Oppie, it was brought up again later, and eventually Milnor did move to the Institute.

Then in 1962, Weil, as always, spent the summer in France. Upon his return in the fall, another mathematician whose office abutted Yang's, told Weil, "You missed a big event. While you were abroad, Lee and Yang had a big altercation."

"What happened?" Weil asked.

"One afternoon, Lee came to Yang's office and they started talking ever louder, until they were outright shouting at each other. They went at it for hours. I couldn't understand what they were saying, it was all in Chinese. In the end Lee stormed out of Yang's office, slammed the door, left Princeton, and went directly back to Columbia."

I remember Weil's final editorial comment — "So much for the wisdom of millennia" delivered in his thick French accent.

16

The Stone Age and the Fermi Era

André Weil had been a professor at the University of Chicago when Lee and Yang were students there. Unlike his colleagues who titled their courses according to the topics they intended to cover: Number Theory, Differential Geometry, Topology, ... Weil taught whatever interested him at the time, but always simply called his course mathematics. He stayed at Chicago for only eleven years but did his most famous work (the brilliant and far-reaching Weil conjectures fundamental in modern number theory) here. Weil was one of the giants of twentieth century mathematics — he revolutionized algebraic geometry and number theory and did superb work in geometry as well. Together with the great geometer S. S. Chern, he did work that had important consequences for the gauge theories of physics. And he realized the relevance of these results for physics at a very early stage. Frank Yang told me that during his student days he was having lunch with Fermi and his group in Hutchison Commons when Weil came by and told Fermi of these results and his conviction that they must be relevant for physics. Fermi listened politely and when Weil left, shrugged his shoulders, as if to say that guy must be crazy. Weil was evidently too far ahead of his time.

It was a remarkable epoch at the University of Chicago. Enrico Fermi, Robert Mullikan, the pioneer of Quantum chemistry, Edward Teller, Gregor Wentzel, Subramanyan Chandrasekhar were in the physics Department, and statistical physicist Joe Mayer, Willard Libby, the inventor of the method of carbon dating, and Harold Urey, the dicoverer of deuterium, were all in chemistry. Joe's wife Maria Goeppert Mayer was on the research staff in the Research Institute and not on the faculty because of stringent nepotism rules, though she later received the Nobel Prize for the shell model of nuclei which she developed in Chicago; her treatment is often, and justly so, held up as a case in point for feminism. Junior physics faculty included "young Turks" like Murray Gell-Mann and Valentine Telegdi. The mathematics Department was in its legendary Stone Age, so called for its chairman Marshall Stone — the son of the Supreme Court Associate Justice who brought a dream team to Chicago. It was he who had recruited André Weil as well as S. S. Chern; Saunders MacLane, the father of category theory and Antoni Zygmund, the master of Fourier analysis.

Shortly after his arrival in Chicago, Weil was asked to give a colloquium talk. With Stone sitting in the first row, Weil began

"There are three types of department chairmen. A bad chairman will only recruit faculty worse than he, thus leading to the gradual degeneration of his department. A better chairman will settle for faculty roughly of the same caliber as himself, leading to a preservation of the quality of the department. Finally, a good chairman will only hire people better than himself, leading to a constant improvement of his department. I am very pleased to be at Chicago, which has a *very good* chairman."

Stone laughed it off; he did not take offense.

While in Chicago, André Weil was deeply involved in the influential Bourbaki group which he had founded with some of his French colleagues. This group set themselves the monumental task of writing a full modern treatise encompassing all of mathematics. Given the unity they wished to maintain, rather than having the author of each volume identified, they decided to publish the many volumes under the name of one fictitious author. For this fictitious author they chose the name Nicholas Bourbaki after an obscure nineteenth century French General Charles Denis Sauter Bourbaki who, after distinguishing himself in battles in Africa and in the Crimean War, was sent to lift the siege of Belfort during the Franco-Prussian War. Though given plenty of troops and equipment, Bourbaki lost the battle. Then, out of shame, he took his pistol and aimed at his own head, but missed. For better or worse, this Bourbaki series set the style in postwar mathematics. To this day some of the volumes are widely used, though their true authors have never been officially revealed.

Since most of the founders of the Bourbaki group were either in Chicago or in Nancy, France, they appointed this fictitious mathematician to a chair at the fictitious University of Nancago and elected him to a membership in the equally fictitious Poldavian Academy.

While Bourbaki was well ensconced in the Chicago mathematics Department, the physics Department was dominated by Enrico Fermi, the preeminent twentieth century physicist to make major breakthroughs in all three areas of theoretical, experimental, and applied physics. He was a legendary teacher, and his course notes make good reading to this day. His presence attracted not only other first-rate physicists to the faculty but an array of graduate students not seen since Heisenberg, Pauli, Bethe, Wentzel and others were studying with Arnold Sommerfeld in Munich. The theory students included T. D. Lee, Frank Yang,

Geoff Chew, Murph Goldberger, Lincoln Wolfenstein, Sam Treiman and Charlie Goebel. The experimental students included Dick Garwin, Jack Steinberger, Jim Cronin, Jerry Friedman, and Owen Chamberlain. Of the twelve students I just named who all went on to great careers, six have been awarded the Nobel Prize. Chicago became unquestionably the center of the Physics world.

Subramanyan Chandrasekhar, who lived in Wisconsin near Yerkes Observatory, was once scheduled to teach a course to which very few people registered. After a few lectures only two students were left. The drive to campus was not easy in those days before expressways, so the department chairman suggested that Chandra need not inconvenience himself for just two students and if he wished the Department could cancel the course. But Chandra decided to go on with the course, because he found these two students truly outstanding and he greatly enjoyed teaching them. The two students were Lee and Yang and this was the only course ever taught at Chicago, or anywhere else for that matter, in which the *whole class and the instructor* went on to win Nobel Prizes.

It was an era of great discoveries. During his short time in Chicago, Murray Gell-Mann introduced the concept of strangeness. This was the first step on the road that ultimately led him to quarks as building blocks of matter. Fermi, with the Manhattan Project behind him, readily obtained funding for building the Chicago cyclotron, on which he performed the scattering experiments that largely rang in what was to become the golden age of high energy physics.

The era had its lighter moments, too. Feynman came to give a seminar, and in the evening there was a party in his honor at the home of Gregor Wentzel. In addition to all his other talents, Feynman had a reputation as a safecracker. During the war he had repeatedly played the practical joke of leaving his signature in the Los Alamos safe where the most secret documents were kept. Wentzel asked Feynman whether he would be willing to give a demonstration. He gave him his small suitcase, a leather *nécessaire* with a combination lock, and asked Feynman to open it. Feynman took a few minutes and the suitcase was open. Everybody applauded. Fermi then asked Wentzel to lock the suitcase again so he too could try to open it. "He spent the rest of the evening fiddling with it, without success," Gregor recalled "Had Enrico but remembered my birthday…"

Fermi developed cancer and passed away rather fast. He knew he was dying but was remarkably serene to the end, with the satisfaction of someone who in his life had done what he had wanted. Chandra came to see him for the last time and was ill at ease. What do you say to a man who is quickly dying and knows it? But Fermi defused the situation with a joke "Chandra, next time do you think I will return as an elephant?"

Heisenberg came to see him as well. They were the same age and were bound by a deep sense of mutual admiration, which even the war could not destroy. Gell-Mann and Yang also came to Fermi's deathbed, and as they were leaving he reminded them "Now it is your turn." The passing of the baton had been accomplished.

The Serene Sensei

Yoichiro Nambu arrived in Chicago in the last years of the Fermi era. An unassuming man of a fatalistic calm verging on serenity, Nambu's contributions to physics were destined to be seminal.

The major theme already sounded when he arrived on the scene was that of symmetries of the laws of nature. Just as the laws of nature do not change from one time to another, as I already mentioned, they do not change from place to place either, and they are the same in all directions. These directions can be the usual ones we see in ordinary space, or they could be directions in some abstract "internal space" defined by the various fields in physics. This property of the laws of nature of not changing under some transformation (e.g. rotation or translation) is called a symmetry of these laws. There are two basic consequences of such a symmetry, a conservation law (remember energy conservation) and a degeneracy in the particle spectrum. It sounds rather medical, but what is meant by such a degeneracy is that most particles must have partners of *exactly* the same mass as their own. For instance if a certain abstract rotation called isospin is a symmetry, then the electrically positively charged proton must have a partner which is electrically neutral and which has the same mass as the proton. Experimentally this is very close to what is observed, for the neutron fills the bill, except that strictly speaking the neutron's mass is not *exactly* the same as the proton's but rather the neutron is heavier by about one fifth of one percent. The symmetry is not exact! Why would nature tease us with the prospect of a symmetry, and then in the last moment decide to break this symmetry, even if the breaking is so small? It's as if nature were denting its own finely polished laws. But if they can be dented, then why polish them so carefully in the first place?

It is a deep question in need of a deep answer, which was ultimately provided by Nambu and then generalized by Jeffrey Goldstone. Basically, their idea is a quantitative theory of Buridan's donkey. Buridan, a fourteenth-century French philosopher assumed the existence of a perfectly left-right symmetric donkey placed exactly midway between two identical piles of hay. Though very hungry, the donkey, cannot find a rational way to make up his mind which pile to go to. Being French, this donkey needs a rational way to reach a decision. Being a donkey, it is obstinate enough to starve itself to death rather than do anything irrational. But imagine the slightest perturbation of this donkey, like shaking off a fly, and the decision would be made; the donkey would be released from its fatally symmetric situation by what may be called *spontaneous symmetry breaking*. The crucial observation made by Nambu and Goldstone is that if in physics one *spontaneously* breaks symmetries, this leads to the appearance of a predictable number of massless particles, the so-called Nambu-Goldstone bosons. Although an analogy with ferromagnetism — where the similar phenomenon of spin waves is known to occur — was made by Heisenberg in his unified theory, he had overlooked this one redeeming feature of that theory. It remained for Nambu and Goldstone to make this major discovery and it was then found that nature makes ample use of spontaneous symmetry breaking. Without it the extremely successful, so-called *standard model,* in terms of which we have come to think of particles and their interactions, would be unthinkable. Nambu is also one of the founding fathers of string theory and has made a number of other important discoveries.

Nambu is a second-generation member of the superb Japanese school of theoretical physics. The first generation consisted of Yoshio Nishina, Shoichi Sakata, Hideki Yukawa, and Sin-Itiro Tomonaga, the last two of whom were awarded the Nobel Prize. During World War II Nambu was drafted into the Imperial Army. In Japan as everywhere else, the military services were in an intense competition. Radar, one of the crucial technologies of the war, was being developed for the Japanese navy by a group headed by Tomonaga. When the army got wind of this, they sent the young soldier Nambu to find out what Tomonaga was up to, in other words to spy on Tomonaga. I have no idea what good this did for the Japanese war effort, but there can be no doubt that this brought Nambu under Tomonaga's spell and after the war we find Nambu very close to Tomonaga's group.

Tomonaga had just developed quantum electrodynamics, for which he was to share the Nobel Prize with Richard Feynman and Julian Schwinger, each of whom independently developed the same theory in the States. Unlike its American counterpart, Japanese physics was patterned on the highly hierarchical German model. The professor was expected to provide the main theory, but he was not expected to dirty his hands with detailed calculations. Those were left for his assistants. One of the crucial calculations for Tomonaga's new quantum electrodynamics was assigned to Ziro Koba, the most senior of the assistants.

Koba took a long time and when he was asked to report on the status of his work, it became clear that he had made a major conceptual mistake and the problem was reassigned. Koba had to atone for his mistake by shaving his head for one year.

At this time no American physics journals were allowed into Japan by order of General Douglas MacArthur. Thus the Tomonaga group was in the dark about the status of Feynman's and Schwinger's work. But General MacArthur allowed, indeed encouraged, the reading of TIME magazine in Japan. Though not formally a member of Tomonaga's group, but having been initiated in Tomonaga's new theory, Nambu immediately saw that it could be used to calculate an anomalous contribution to the electron's magnetic moment. This, as it turned out, was one of the crucial predictions of the new theory, but just when Nambu finished his calculation, he learned from TIME magazine that he had been scooped: TIME reported on Schwinger's work and even gave his famous "alpha over two pi" formula. "It agreed with my result," is how Yoichiro modestly put it to me.

To understand the way Yoichiro talks about physics, you have to factor in this modesty and keep in mind the fact that at some level he is a magician. He has had many brilliant ideas and I remember discussions at which he would talk about them for the first time. Invariably, he would start with a flood of well-known results, of pretty equations and in his beautiful handwriting get some famous names on the blackboard as if to say "I'm just making a small observation which probably isn't even new." For those who don't know him, this would be the right moment to tune out. With Yoichiro the rewards come later, when he finally gets to his own work: then he suddenly pulls a whole array of rabbits out of his hat and, before you know it, the rabbits reassemble in an entirely novel formation and by

God, they balance the impossible on their fluffy cottontails. When he's done, you can ask where these rabbits came from, and a meticulously logical explanation will be provided, but you will never find out the full story of these rabbits, where they were before they showed up on the blackboard, what they did before staging their appearance. In this respect, Nambu's style closely resembles that of Feynman and Dirac — his extremely original thought processes, though leading to very deep and perfectly logical results, remain clouded in mystery.

Although he has spent most of his life in the U.S., Nambu remains very Japanese in his etiquette. In Japan you try as much as possible to avoid negative answers. If someone asks you whether the number 2 is larger than the number 3, you can of course answer with an emphatic no, or let me just tone that down a notch, to a polite no. But if someone asks you whether you want to join them for dinner, whether you are free or not, whether you have eaten already or not, you do not answer in the negative, you say something like "Isn't that a lovely idea?" thus avoiding an answer. A positive answer avoided becomes a negative answer that does not hurt anyone's feelings. Nambu's way of complying with this rule is to answer any question with a yes, but with a pause before that word. The longer the pause, the stronger the no becomes. His absolute no is defined only asymptotically as a yes delayed for an infinite amount of time.

At one point Nambu was appointed chairman of our department, even though he took a very long pause before accepting the dean's invitation. During the first year of his chairmanship he became extremely popular. Faculty members began approaching him with the most outlandish requests — wanting to double their lab space overnight, or to undertake major renovations. People would tell me that in Nambu we had the best chairman ever, and as an example they would relate a request of theirs, which, unlike his predecessor, he had approved. I would ask them whether he had said yes right away "*no,* he took a few minutes to think it over and then said yes, he knows what he is doing." He sure does. Then a year later these same people would tell me that Nambu is the worst chairman we ever had, "A year ago he promised me this and that and he didn't do a thing about it." I would then instruct my colleagues in the finer points of Nambu's use of the word yes.

Nambu has a number of hobbies. He built himself a telescope and with it he discovered a new comet. He reported his discovery to wherever you are supposed to report such discoveries, but alas there is no comet Nambu; some other amateur beat him to the punch. He also has taken some marvelous photographs. He has a candid shot of Albert Einstein taken from a car in Princeton as the disheveled old man walked towards it to accept an offered ride. Yoichiro had met Einstein in Princeton in spite of a prohibition imposed by the Institute's director, J. Robert Oppenheimer, Oppie as everyone called him. When the new members arrived in the fall, Oppie, in his characteristic manner announced that though Einstein was working there, he should not be disturbed: no one was to see him without first obtaining Oppie's permission. Nambu knew Bruria Kaufmann, Einstein's assistant, and managed to arrange a meeting. They had a pleasant discussion of physics; Einstein had felt quite lonely and was happy that a young man had finally come to visit him.

Oppenheimer, Hero
or Antihero?

J. Robert Oppenheimer was very different from the legend. He was still in Princeton when I was there for the 1964–65 academic year, and I got to see him in action.

In the days before not only the internet but even the Xerox machine and the circulation of papers were not as quick as they are now. When you wrote a paper, you had some twenty mimeographed copies made, and you would send these "preprints" to the physicists you believed to be the most interested in the paper and, as a courtesy, to certain prominent people in the field. To read a preprint in those days was a much more personal matter than it is today. There were no word processors, the equations were written in by hand, and you related directly to the author's handwriting. You got to see maybe three papers a week, compared to some twenty-five new postings on each internet archive on an average day these days, which adds up to fifty a day if you follow two archives. It was a much more leisurely, though not less intense, business and you depended much more on going to seminars and conferences.

The theory seminar at the Institute was run by Oppie himself. He got all the preprints, read some of them, and then invited the speakers as he saw fit. During a seminar, Oppie, puffing away on his pipe, would stop the speaker to ask the kind of question to which the answer was invariably "That is a very good question — I was just getting to that point." From the way he asked his questions, it was clear that he, having read the paper, had prepared this question and was inadvertently acting as the speaker's straight man. After receiving his "That is a very good question ...." compliment, Oppie would turn around from his first row seat with the demeanor of a pianist receiving applause for having performed a technically difficult piece. It became both predictable and embarrassing. Many in the audience just looked away to avoid eye contact with Oppie. But once, he got his comeuppance.

The speaker was Kerson Huang from MIT. Kerson introduced a parameter and Oppie interrupted:

"Kerson, what about the sign of this parameter?"

"That is a very good question." Oppie took his bow and Kerson went on "Its sign must be either plus or minus" — a polite way of saying it is totally irrelevant.

Ten minutes later Kerson introduced a second parameter and emphasized that it must be positive, its sign can never be minus, making it clear that Oppie had asked his prepared question in the wrong place. The people could barely suppress their laughter.

Sad to say, this occurrence was far from atypical. When I first arrived at the Institute, I was very pleased to be invited to Oppie's office for a conversation. When I showed up, he told me that he had just received a paper on superconductivity which he found very interesting and wanted to tell me about. The study of superconductivity fits in the field of what in those days was called solid-state physics. I am a particle physicist, but unbeknownst to Oppie, I was very interested in superconductivity at that time. Taking his interest in superconductivity at face value, I started asking him some questions about what he was telling me. He could not answer any of them, and it became clear that he did not fully understand the paper he was talking about. He must have realized this, for he became visibly annoyed. I guess he expected me not to know anything outside my field, and to be awed by his breadth. I was stunned, and a few days later mentioned this to Sam Treiman, a Chicago graduate, by then a professor at Princeton University. "That Oppie is superficial is the worst-kept secret around here," Sam explained with his characteristic sense of humor.

But the worst was yet to come. Shortly before Christmas, a seminar was to be given by Wally Greenberg, who had just made the case for what later became known as color, an essential quark attribute. The brilliant talk was poorly attended, most people having already left town for the winter break. Of the ten people or so in the audience one was the professor of theoretical physics at a neighboring university, who had made the drive to Princeton just for Wally's talk. The seminar room was small, and the first row was taken by a few heavy leather armchairs usually occupied by the senior members of the Institute: Oppie, Frank Yang, and Freeman Dyson — all, except Oppie out of town. So the visiting physicist just plopped down in one of the armchairs. When Oppie made his entrance, he headed directly for this man and loudly reprimanded him: "Get up, this seat is not for you!"

Discussion of Oppie's behavior abounded among the young physicists then at the Institute. What caused all this arrogance? A number of explanations seemed possible. By the time I met him, Oppie's security clearance had been revoked, in the aftermath of his opposition to the construction of an H-bomb, in view of fears expressed by some that this might have been politically motivated. It was possible that this had caused such trauma as to bring to life a different Oppie, bitter and capricious.

Another explanation started from the fact that Oppie's greatest achievement had been his stewardship of the Manhattan Project, in which his role had been mostly administrative. He was conceivably *the* greatest science administrator of all time, appointed in his late thirties as boss of the likes of Enrico Fermi, Eugene Wigner, Hans Bethe, Richard Feynman and Edward Teller. Oppie's career as a creative scientist and as a teacher (Bob Serber, David Bohm, Hartland Snyder and Bob Christy had been among his students) came to an end with this appointment. He was best known for three pieces of work.

With his teacher, the Nobel Prize laureate Max Born, he devised a particularly useful approximation scheme in the quantum theory of molecules.

Somewhat later, he pointed out that for experimental reasons the proton could not be the positively charged particle predicted by the Dirac equation. But, in the end it was Dirac and not Oppie who came up with the revolutionary concept of antimatter, predicting the existence of a new, never before seen particle, which shares all the attributes of the electron, except for the fact that the sign of its electric charge is the opposite. This new particle, the positron, was soon thereafter experimentally discovered by Carl Anderson at Caltech.

Then, with his talented student Hartland Snyder, Oppie did some early work on black holes.

These all are important papers, but hardly in the same league as the work of the people whose boss he had become, let alone that of Einstein, whose boss at the Institute he was yet to become.

Even in the Manhattan Project, the seminal ideas stemmed from others. Oppie was certainly savvy enough to understand the disparity between his exalted office and his status in the field. Not that any of his "underlings" ever rubbed this in — they didn't need to. Anyway, none of them was into administration. Yet after the war, it was Oppie who was the one lionized. In his heart of hearts he must have known very well that this glory was a temporary, transient phenomenon. And then came the security clearance debacle.

In 1965, Oppie was fighting tooth and nail to prevent Heinar Kipphardt's play about him to make it to Broadway. That play was not particularly sympathetic to him, and it delved into some of his character problems. He was not detached enough to realize that his role in history was assured — not through his contributions to physics but through the remarkable drama attached to his persona. To paraphrase Feynman, who had once told me that Schrödinger had been "the Harry Truman of physics: not necessarily the greatest, but when his hour came he lived up to it," I would say that Oppie was the Mary Stuart of physics, a tragic, in posterity's eye eternally controversial figure, now a hero, now just the opposite.

Viennese Physics,
Then and Now

No one can deny that an immense scientific-technological undertaking like the Manhattan Project or the Human Genome Project has to be centrally coordinated lest it lose focus and get mired down. Oppenheimer was the first, and to this day the greatest administrator of such a huge enterprise. His role as the director of a major research institution, the Institute of Advanced Study in Princeton, is more controversial. For research to flourish on the scale of a single university or a single institute, the presence of a strong administrator at the top is not always a blessing, and often it is not even necessary.

One classical example is provided by Vienna in the last decades of the nineteenth century. Although when one thinks of science in the German-speaking world, cities like Berlin, Munich, Göttingen and Zurich first come to mind, there is no denying that at the end of the nineteenth century, Vienna is where it was at. That is the period in which statistical mechanics and the related kinetic theory of gases were created, and though Josiah Willard Gibbs at Yale and the Scot James Clerk Maxwell (the great genius who discovered the laws of electrodynamics) also had a hand in this, the dominant figure in this enterprise was Ludwig Boltzmann in Vienna. He had studied there with Jozef Stefan, whose work on black-body radiation is still of fundamental importance. Ernst Mach, whose rethinking of classical mechanics drove Einstein on his quest for general relativity, and who is now widely known in connection with supersonic flight as well as for his contributions to psychology and positivistic philosophy, was also in Vienna. So was Johann Loschmidt, who first correctly determined the size of the then-still-hypothetical molecules and independently found what we now call Avogadro's number. Austria was a big power, an empire on its way down. During the fall of an empire, just as during its rise, the arts and sciences flourish, just as long as this fall is swift enough and doesn't get dragged out. If it lacks gradient and gets dragged out over many centuries, as was the case, say, with the Western Roman Empire and the Arab Caliphate, then instead of a flourishing culture, dark ages of religious fanaticism and obscurantism follow. These can then only end when the imperial fantasies are abandoned. So it would be reasonable to say that it was the fall of the Habsburg empire that brought this tremendous constellation of physicists to its premier institution of higher learning. It was a loosely knit bunch that didn't see eye to eye on many issues. Mach and Boltzmann, for example, were major antagonists. Yet it all came to pass without any Oppie minding the store.

The glory days of Vienna didn't last long however. Was it the fall of the Habsburg Empire or the death of the protagonists? It is hard to tell. Ultimately, like empires, great scientific organizations also have a finite lifetime. The fact is, much of the next generation went abroad to work — Schrödinger and Pauli for example, the latter even went to study in Munich. But Vienna continued to house some first rate physicists. Hans Thirring, in a remarkable set of papers, showed how Mach's ideas were accommodated in Einstein's general relativity. This is the so-called Thirring-Lense effect, with verifiable experimental consequences. Unfortunately, for existing celestial bodies this effect is so small that it cannot be experimentally detected. Then around 1960 the Russian theoretical physicist V.L. Ginzburg made the suggestion that by shooting a precision gyroscope into orbit around the earth, the Thirring-Lense effect could be observed.

I was a student in Vienna when this suggestion was made, and I asked Thirring what he thought of it. He was not at all happy. By then he was in his seventies and, though he still came to the University every now and then, he had retired and had been elected a member of parliament. He told me that the Thirring-Lense effect was the best thing he had ever done in physics. He even had a letter from Einstein complimenting him on having found the most beautiful effect in all of general relativity. "And now they want to test it. So say some Russian or other shoots a gyroscope in orbit. Most likely such a Russian-made gyroscope will malfunction and the effect will not be found. It will be announced with great hoopla and I will die an unhappy man. Then later, some American will redo the experiment properly and the effect will be seen, but by that time I will already be dead. At my age do I need all this aggravation? Let them go at it after I am dead."

Hans Thirring needn't have worried; the Russians never attempted the experiment, and when Leonard Schiff at Stanford later made the same suggestion as Ginzburg, his colleague Bill Fairbank, a master experimentalist, started building the requisite gyroscope. It took a very long time and by the time both Hans Thirring and Bill Fairbank had passed away, the experiment had still not been performed. Finally, in 2007 a NASA experiment confirmed this "most beautiful effect in all of general relativity."

Hans Thirring was a small, bespectacled man who wore his sense of humor on his expressive face. As a member of the Austrian Parliament at the height of the Cold War, at the time of the Cuban missile crisis, he came up with a truly innovative idea. He saw disarmament as the only way to avoid a nuclear war between the Russians and the West. But disarmament is fraught with obvious social problems, chief among them how to reintegrate into productive society the career-military once they are no longer needed, and how to finance this reintegration. To find out how this could be done, he proposed that Austria voluntarily and unilaterally disarm and offer itself as a case study to the United Nations. At that point Austria maintained only limited military forces, one of the main purposes of which was to stall the advancing Russians long enough for the Austrian president to put in a phone call to Washington. Thirring suggested that before disarming, Austria should sign mutual non-aggression treaties with all her neighbors. If a disarmed, defenseless Austria is then nevertheless attacked and conquered by one of these neighbors, then this will have proved that disarmament is unfeasible; World War III is unavoidable, and the only drawback will have been that as in the days of Hitler, Austria will have fallen one year before hell breaks loose. By then a seasoned politician, Thirring did not fail to point out that the disarmament observers to be sent by all UN member states would generate tourist dollars for the local economy. As could be predicted, the next time Thirring was up for reelection, the whole Austrian military brass came out in force to campaign against him, and his political career was over.

At the University, Hans Thirring was succeeded by his son Walter Thirring, Austria's leading postwar theoretical physicist, the discoverer of the famous quantum field theory now called the Thirring model. Its importance stems from the fact that this theory can be exactly solved. Various questions that cannot be answered in the case of important but much more complicated theories can be answered simply for the Thirring model. These simple answers can then be used to guide work on the more complicated theories. I am particularly indebted to Walter Thirring. He was my teacher.

The scientific life presided over by Walter Thirring was very exciting. We had long visits from Bruno Zumino, who gave a course in his Italian-lilted German, from Bob Karplus, with whom I collaborated on two papers, from Reinhard Oehme my current Chicago colleague, from Edward Teller's former Chicago student Boris Jacobson in Seattle, from the University of Maryland solid state physicist Richard Ferrell, from Jeremy Bernstein, the *New Yorker* writer. Then we had shorter visits from Res Jost (Zurich), Joseph Maria Jauch (Geneva).

I vividly recall a visit from P.A.M. Dirac. Thirring was at an important faculty meeting and asked me to take care of Dirac. I was truly awed, and even Dirac, a very tall and extremely thin man whose social skills were usually limited, sensed this and started giving me a little lecture. We were in what was officially still the office of Schrödinger, who was already on his deathbed. Dirac took this as his cue and told me that in theoretical physics there are two major styles of research. On the one hand, you can closely follow what the experimentalists are doing and when a contradiction arises, remove it through a new theory. To his mind, Heisenberg was the most brilliant practitioner of this approach. On the other hand, Dirac went on, saying that you can closely follow the theories and when it gets to a contradiction in the accepted theoretical framework, you construct a new and more powerful theory. Here he singled out Schrödinger as the most brilliant practitioner. Most would agree that in reality this distinction fell on Dirac himself.

As he was going through this lecture, others began entering the room, including Jeremy Bernstein who, launched into an informal interview of Dirac. First he asked Dirac whether he pronounced his name the English way (dee-rack), or whether he preferred the French pronunciation, with the vowel a pronounced as in say the word "rah". "Oh, I really don't hear any difference" was Dirac's answer. He asked what the great old man thought of the future of physics. Dirac was very worried that quantum field theory, of which he was a cofounder, seemed to yield infinite values for certain quantities. Normally these infinities are removed by what is called *renormalization*, and by now, it has been developed into a mathematically rigorous procedure that in the end yields finite and consistent results. To the end of his days Dirac forcefully disapproved of this clever procedure:

"For centuries mathematicians have found it useful to neglect the infinitely small," he said, "As of late the physicists have suggested that one neglect the infinitely large as well." But if not renormalized quantum field theory, then what? "Nonassociative algebras, of course" was Dirac's cryptic answer. To this day I have no idea what he might have had in mind.

Physicists from the Austrian Diaspora also showed up at the University of Vienna. Marietta Blau, who with Herta Wambacher pioneered the use of the emulsion technique, and whose career was disrupted by the Nazis, came by for a visit. She had spent the war in the States but had failed, until recently, to garner the recognition she was due. History however has a way of making amends.

Two visitors in particular were of great interest to me. One was Guido Beck and the other Bruno Touschek.

———————

Touschek, an Austrian without enough Aryan blood to make it smoothly through the Nazi era, had left after the war for Italy. But, in Italy in those days, to teach at a university, you had to be an Italian citizen.

Touschek's father, a high ranking officer in the Austrian Navy during the First World War, would not hear about his son becoming a naturalized citizen of an "enemy nation." So Touschek had to content himself with a pair of research jobs that earned him an acceptable income and even allowed him to supervise students, the brilliant Nicola Cabibbo among them. In physics Touschek had introduced the important concept of chiral symmetry and produced the theory of beam stability for particle colliders, a theory still in extensive use. It would be correct to say that he invented electron-positron colliders. He was a very animated, joyous character, partial to a few bottles of quality wine.

On one occasion, happily disposed by such consumption, he was stopped on the streets of Rome in the wee hours of the morning for disturbing the peace. As their counterparts everywhere else would have done, the Roman policemen asked Touschek what he did for a living. "I study time reversal." He meant one of the important symmetry operations in physics, whereby the flow of time is reversed. The policemen, not privy to the scientific value of this problem immediately arrested Touschek, assuming that by turning time around he wanted to bring back Fascism, which was against the law.

Touschek invited me to give a talk in Rome on what was to become my doctoral thesis. But when I got to Rome, the whole university was on strike. The entire country was in political turmoil during the last days of the Tambroni government, which without any time reversal, had started in earnest going about the restoration of Fascism. I was put up in a Roman *pensione* and told to do some sightseeing till the reopening of the university. One day Touschek invited me to go snorkeling in the Lago Albano, just under the Castel Gandolfo, the pope's summer residence. Shortly after one o'clock he announced that we had to hurry back into the city because he had to join a protest demonstration at five. The drive back was maybe half an hour, so I couldn't understand why the big hurry, especially since snorkeling in this pristine lake was so very pleasant in the canicular Roman heat. Touschek explained that he had to go home, shower, shave, change to a good suit with tie and then walk to the university.

"For a *demonstration?* Couldn't you just drive there the way you are?"

"You obviously know nothing about Roman demonstrations. They always end up with the demonstrators building barricades out of whatever cars they can find, then they start throwing stones, and sooner or later they set a few automobiles on fire. Let them use someone else's Fiat 500, not my beautiful Lancia." He did have a beautiful Lancia, the doors of which were hinged at the front and rear of the car so that they both opened in the midsection, the kind of design since discontinued because it had too much class.

"Then I also have to shave and change," he went on, "because Roman policemen are instructed to chase the demonstrators with their rubber clubs. They start by clubbing the bearded and the open-shirted, who in Police manuals are identified as the rabble rousers."

The day after the demonstration, the university reopened, I gave my talk, and shortly thereafter, what else, the Tambroni government fell.

Touschek had contacts in the jet set. When Gina Lollobrigida, the beautiful and intelligent movie star, had to hide from reporters, she stayed at Touschek's place, and it fell upon this theoretical physicist to chase away the paparazzi. He was also very close to that other Austrian expatriate Wolfgang Pauli and had visited him at his deathbed in hospital room 137 — a number of great importance in physics, as it *almost* equals the inverse of Sommerfeld's fine structure constant alpha.*

As was his wont, Pauli often discouraged the young Touschek from whatever physics he was doing by telling him that he was too late, and that whatever he was doing had already been done. When Touschek showed up in room 137, Pauli received him with the words "As always Touschek, you are too late!" No elephants.

The other visitor who fascinated me, Guido Beck, was a very small man with fine white hair surrounding his shiny baldness, always modishly attired and carrying an umbrella or walking stick, which he used eloquently in conversation. He was visiting from Brazil. Before going there in the Thirties he had worked with Heisenberg in Leipzig and Bohr in Copenhagen, and in 1931 had written a paper with Hans Bethe and Wolfgang Riezler, which to this day remains one of the most inventive scientific pranks ever. The great astrophysicist Sir Arthur Eddington had in old age proposed that the inverse $1/\alpha$ of the fine structure constant α is not only close to, but exactly equal to the integer 137. We now know this to be nonsense on the basis of more recent experiments, but even in those days it was totally off the wall. To parody Sir Arthur, and to make a subtle political statement, Beck, Bethe and Riezler wrote a paper, claiming to have found on the basis of Eddington's theory a relation between the position τ (= −273 degrees) of

---

* α is approximately equal to $1/137$ which is a small number and this renders the approach of Feynman, Schwinger and Tomonaga to quantum electrodynamics so successful, in that one can include all contributions proportional to α, to α squared, to α cubed up to some pre-chosen power of alpha, dictated by the accuracy to which one performs the calculation. On account of the small value of alpha, one can be sure that the omitted terms are all smaller than those included in the calculation.

absolute zero on the *arbitrary* Celsius scale of temperatures and the inverse of the fine structure constant α. Their relation reads $\tau = -(2/\alpha - 1)$, which numerically translates into $-273 = -(2/(1/137) - 1)$, and seems to be obeyed. What is wrong is that α (which is not even strictly, but only approximately $1/137$), is a fundamental constant of nature, whereas the position of absolute zero on a manmade temperature scale like the one of Celsius is purely conventional and as such has no fundamental meaning whatsoever. Why Celsius and not Fahrenheit? Their result is not less preposterous than if someone were to claim that he can explain the number 23 on Michael Jordan's Bulls jersey from the fine structure constant since by multiplying this shirt number by the number 6 which is the number of Bulls on the court plus 1, one obtains the number 138, which differs by only 1 from the inverse of the approximate value of the fine structure constant.

Their argument itself is pure parody; they even invoke frozen degrees of freedom, in a not all that veiled allusion to the dawn of the Hitler era. They submitted the paper to *Die Naturwissenschaften,* a German analog of the famous British journal *Nature.* Adolf Berliner, the publisher, a well-meaning amateur, was impressed by the young, but already famous, name of Hans Bethe among the authors and published the paper without delay. Then many letters started arriving in the editor's office asking for further clarifications. Berliner forwarded them to the authors and in lieu of a reply, they admitted that the whole thing had been a harmless prank. In that case, Berliner insisted on a printable apology.

They wrote something to the effect that they had indeed made a prank, and that they were confident that no harm was done, for no one but a total idiot would have been taken in. But the trouble was, the editor had letters from a number of famous people who *had* been taken in,

Max Planck, the discoverer of energy quanta, among them. Calling them idiots would add insult to injury. No retraction was published, but the matter made it into the press and the three authors — Beck was Jewish and Bethe half-Jewish — were crucified in the newspapers as degenerate Jews desecrating the holy German science.

Not surprisingly Beck and Bethe fled to the Americas a couple of years later. Bethe ended up in the U.S. and is now well known. Beck decided to go to Brazil, so he traveled to Lisbon and was to continue by ship to Rio. There was a few months' wait for the next ship, and he went to the ancient university in Coïmbra to while the time away. The physics department asked him to lecture, and he gladly obliged. Then, when his ship was about to leave, he said his good-byes to his hosts, but before he got anywhere near the ship, he was arrested. One of his local colleagues charged that Beck was in possession of a coded list with the names of three hundred Portuguese officials he was sent to assassinate. The list was found in Beck's room.

When Beck was shown this list, he recognized it as the menu — with all the three hundred types of *smørrebrøds* — from Oscar Davidsen's, a legendary Copenhagen restaurant, which he had kept as a cherished souvenir from his days of working with Bohr. Beck protested. The police were willing to give him the benefit of the doubt requesting only that the Portuguese Embassy in Copenhagen obtain a copy of this menu so they could verify his claim. The verification took a few weeks, by which time Beck's ship had left Lisbon harbor. Back to Coïmbra he went and finished his lectures. The next time though, they let him leave.

––––––––––

Beyond these stimulating visits we also had some quite amusing visits of a "different kind." In 1960 a South American monk showed up at the Institute of Theoretical physics in Vienna. He claimed he had the ultimate theory of everything and that his brethren, in a missionary spirit, had sponsored his round-the-world trip so that he can disseminate this knowledge. He wouldn't think of wasting his time telling mere students about it but demanded to be led to the professor, in our case Walter Thirring. Before he would divulge even to Thirring the ultimate truth, there was, like often happens in religious orders, a kind of induction ritual: Thirring had to pass an exam. The monk gave Thirring this exam, and my teacher intentionally flunked it, thereby ridding himself of this visitor without finding out his version of the ultimate truth.

Cranks approach scientists more often than you would guess. The main strategy for dealing with them is never to get into an argument, for they will not spare any of *your* time to convince you that they are right. The other useful trick is to convince the crank that you do not have the required expertise to be initiated into their sacred truth.

Another strategy: Academician N.N. Bogoliubov at Moscow State University was once approached by a crank. "I unfortunately am not qualified to discuss your work" Bogoliubov told the man, "but Academican Lev Landau is working on related problems. He is the man you are looking for." Obviously, the relation between the two academicians was not the friendliest.

On another occasion, while Thirring was away, two men from the Shah of Iran's Vienna Embassy showed up, all agitated, at the Institute. It fell upon Thirring's assistant Kurt Baumann and me to take care of them. One of the two Iranians excitedly told us about their discovery, while his companion reverently nodded along. They had apparently discovered that time does not exist. Their proof was eminently simple: "By the time I say now, now is already over. Quod Erat Demonstrandum." Did we not agree that this was a major discovery, and what should they do with it? Armed with our two principles, we agreed that the discovery was major and suggested they work out all its implications, especially practical applications, for they would likely make a great deal of money. This did the trick; the two left happy men.

20

The Importance of a Sense of Humor

In March of 1929, strolling through Göttingen on a starlit night, the twenty-six year–old Fritz Houtermans and the woman he would end up marrying twice, were about to engage in the kind of emotion-laden conversation young lovers borrow from the books they have read and the movies they have seen.

"Don't they shine beautifully tonight?" the woman started.

"I've known since yesterday why they shine" Houtermans replied, in defiance of centuries-old conventions of romantic literature. Indeed, he was the first man ever to understand that the light coming to us from the stars is the result of nuclear reactions taking place in them. It was he who coined the term thermonuclear for these reactions.

There remained the most important problem of understanding precisely which nuclear reactions take place, so that one could test this idea. After an inspired, if incomplete, attempt by Carl Friedrich von Weizsäcker, this was fully worked out in a classic 1939 paper by Hans Bethe, who received the 1967 Nobel Prize for it. The first reference given in Bethe's Nobel lecture is to Houtermans' work.

Barely four years after his major insight, Houtermans had to flee Germany for two reasons: he was one quarter Jewish, and he was a member of the Communist Party. He left for England with Charlotte, the woman from the Göttingen stroll and by then his wife. He ended up working at the *His Master's Voice* laboratories outside of London. There he almost discovered the laser; he had the correct idea, but not the needed experimental virtuosity. His work failed. Add to this his hatred of English food, and in 1935, Houtermans decided to accept an appointment at the Kharkov Physico-Technical Institute in the Ukraine. In those days Kharkov was the premier place in Soviet physics, L.D. Landau was professor there. Houtermans started a successful collaboration with Valentin Fomin, but in 1937 it all came to an end as the wave of terror of the Stalinist purges engulfed the Kharkov Institute. Houtermans was arrested and Fomin "escaped" arrest by committing suicide. The NKVD tortured Houtermans. They made him stand four or five feet from a wall and lean against it with only the tips of his fingers touching the wall, which after a short while became enormously painful. No wonder he confessed to being a German spy. He gave them detailed drawings of a contraption he claimed to have built to measure the speed of Soviet planes flying overhead. Then he told his interrogators that he had passed on the results of his measurements to two German officers whom he endowed with names borrowed from the history of the Napoleonic Wars.

During his years in the Gulag, Houtermans kept his sanity thinking about number theory, "the only experimental science that you can do without a laboratory." After the war Emmy Noether's student Bartel van der Waerden, suggested to Houtermans that he publish his number-theoretic results, but he did not. In the Gulag he ran across Fritz Noether, Emmy's brother, himself a distinguished mathematician, who on account of his communist leanings, chose to go to the Soviet Union rather than follow his sister to the States. He paid dearly for this decision. Under torture, he confessed to having spied on the Soviet navy and drew a 25-year sentence.

Charlotte Houtermans and the couple's two children managed to get out of the USSR and made it to the States, whence she organized a campaign for petitioning Beria and Stalin for her husband's release.

In 1940, in the wake of the Molotov-Ribbentrop pact between Germany and the USSR, Houtermans, technically a citizen of the Third Reich, was handed over to the Gestapo. From there he was freed three months later through the courageous intervention of Max von Laue, the Nobel Prize laureate. In 1941, Von Laue also got Houtermans a research job at a private laboratory headed by Manfred von Ardenne, a wealthy scientific self-promoter, who was going to build atomic bombs for the Post Office, and who at war's end would parlay his "expertise" into one of the top scientific jobs in the communist East. Even in this environment, Houtermans managed to discover the role of Plutonium in the construction of nuclear weapons. To his credit, though, he managed to send a message to Fermi in America about these developments "Hurry up! We are on the track!"

In 1945, Manfred von Ardenne fired Houtermans in the wake of the "tobacco scandal." Tobacco was very scarce in Germany in those days, and Houtermans, a chain-smoker, sent an official letter to a leading cigarette manufacturer demanding Macedonian tobacco, from which he claimed he could extract heavy water needed in his war work. He got the tobacco, smoked it and then even got a second batch, but this led to an investigation, which resulted in his dismissal. For the remaining months of the war he managed to relocate in Göttingen, where they created some kind of a job for him and where at war's end he would walk the streets picking up the cigarette butts discarded by American soldiers.

As things started returning to normal, he was asked by an administrator in Göttingen, "Excuse me, sir, who exactly are you, and what are you doing here?"

"You see," Houtermans replied, "I can best explain this to you by telling you the story of the little old Jew who went every day to his favorite tobacconist's, bought his tobacco and then lingered in the store for a while. After about a year, the tobacconist asked him

'Excuse me, sir, but who exactly are you, and what are you doing here?'

To this the little old Jew replied, 'Don't you know? I am the little old Jew who comes in your store every day to buy his tobacco and then lingers for a while.'"

Not only was tobacco scarce in Germany in the last days of the war, but so was fuel, which prompted Houtermans to propose that all that fuel needed to cross the Channel could be saved if henceforth British planes were to bomb Britain and German planes to bomb Germany.

For the food shortages in the early postwar years Houtermans, a connoisseur of the Carbon cycle that makes stars shine, also had a simple solution. There was also a blood shortage, and the blood banks were trying to entice donors with the promise of a nice slice of blood sausage. So Houtermans came up with the blood sausage cycle: A donates blood to B and gets a sausage; then B donates blood to C and also gets a sausage; finally C donates blood to A and gets his sausage as well. No blood will have been lost by A, B, and C and they will all have gotten to eat a nutritious slice of sausage.

Yes, the man had a keen sense of humor. Though born on the outskirts of the then German Danzig, Houtermans grew up in Vienna, and had fully mastered the biting and bitter humor for which this city is famous. With the drama life had dealt him, this sense of humor served him well; it became his life preserver.

And then there are the marriages to Charlotte, herself a physics PhD from Göttingen. In 1930 they were attending a conference in Odessa and after it ended they took a trip to the Caucasus. There, with Wolfgang Pauli as one of the witnesses — yet another role for Pauli — they tied the knot. Upon their return to Berlin they were at the center of the physics life there. They gave frequent "*Eine Kleine Nachtphysik*" parties, where physics, the arts and politics were discussed, and lots of liquor was consumed. During the war the spouses were on different continents and the marriage unraveled. Houtermans married another woman, who then left him for the distinguished nuclear theorist Otto Haxel. Then in 1953 Fritz and Charlotte remarried. It didn't last and Fritz Houtermans was to go through yet a fourth marriage.

In 1953 he was appointed professor at the University of Berne in Switzerland, a somewhat provincial town as such things go. He modernized the laboratory there and managed to attract a number of first-rate young people, especially Walter Thirring.

Joseph Fouché, Napoleon's minister of police, lived out the last years of his eventful life in Linz, which in German, as Stefan Zweig, pointed out, rhymes with "Provinz." In this sense of the contrast between the road and the destination, Berne was Houtermans' Linz. He managed to scandalize even this sleepy town with his driving, to the point where his license was revoked and a special license, valid for only one particular motorcycle, was issued to him. He didn't keep this vehicle in good repair, and after a while it could no longer turn left, only right. When he invited speakers, he would want to show them the way to their hotel. He'd ask them to follow him in their cars. The hotel was a few blocks from the university, but with the constraint of no left turns, this would invariably end up a very long drive.

Besides the pleasures of lingering at fine tobacconists, there is an underside to chain-smoking — the lungs get destroyed. Even being the first to understand that nuclear reactions fuel the stars is no protection. Houtermans' lungs went out of service in 1966.

The Russian Style

The Communists in the Soviet Union and its satellites and the Nazis in Germany have inflicted major suffering and hardships on the scientists unfortunate enough to be ruled by them. Of these dictatorships, only the Soviet Union — even though it also collapsed in the end — has lasted long enough to have considerably affected the history of physics. Remarkably, the more than seventy–year lifespan of the Soviet Union was sufficient to lead to the formation of a distinctly Russian style in physics. Though not intended by the Soviet rulers, this was a genuinely beneficial development.

The crucial role was played by Lev Davidovich Landau — Dau to his colleagues. He towered over the whole Soviet physics establishment to such an extent that his stylistic imprimatur is to be seen on all Soviet physics and even on much of Soviet mathematics. Dau was a thoroughly difficult man, as much as a consequence of his quirky personality as of the absurd times in which he lived. He was aware of this and on his office door he posted the sign "Landau. Beware! He bites."

To a large extent, the center of Soviet theoretical physics was wherever Landau happened to be, in the

Ukrainian city of Kharkov in the beginning — where Houtermans joined him — and later in Moscow. He had many brilliant students and with them pursued many directions of research. It can be safely stated that there is hardly a field of physics in which Landau has not made one or more seminal contributions.

Landau had been in Copenhagen in the early days of quantum mechanics and was widely admired worldwide. He apparently had some kind of aversion to writing papers and so he would lecture to his collaborators about what he was doing, and from the notes they took, *they* would write up the paper he would then publish. For his superb treatise on theoretical physics, E.M. Lifshitz was his co-author. It is often said that Landau did not write one word of these many volumes and that Lifshitz did not contribute one idea. That must be an exaggeration, for Lifshitz did some good work of his own.

Landau had a system for classifying theoretical physicists into four categories. The best category was the triangle △, sharp at the top and thus able to come up with major new ideas, while at the same time sturdy at the bottom, master of all the techniques needed to develop the idea and having the stamina to do so. The next category is the diamond ◊, still sharp at the top, creating new ideas, but without the stamina and technical abilities to fully develop these ideas. Next came the square □, not an originator of great ideas, but someone able to develop an idea already in circulation. The worst of all Landau's categories was the upside down triangle ▽ — no ideas, no stamina and no technique. He supplemented this by a more refined numerical scale to grade physicists. There was a special top grade reserved for Einstein. Heisenberg, Dirac and a few others were next to this very top of the classification. Landau promoted himself to this category when he came up with the theory of superfluidity, which ultimately earned him the Nobel Prize.

While he moved to Moscow just in time to escape Stalin's purges in Kharkov Institute, the horror caught up with him in the capital. In 1938, Landau was arrested and imprisoned in the infamous Lyubyanka prison for a full year. The famous story has it that in jail Landau was pacing deep in thought. A guard asked him,

"Why do you keep pacing?"

"Because I am thinking."

"Work! Don't think!" the guard admonished him.

Understandably, this imprisonment was a devastating experience, which may explain why Landau later on did not dare to defy Stalin when it came to the Russian effort to produce nuclear weapons.

After the war he was no longer allowed to travel to the West. They could have kept his wife hostage in the Soviet Union, but this, it was believed, would have only given him one more incentive to defect. After the severe brain damage he suffered in a 1962 car accident, he could no longer read and he would sign his name on documents put before him supposedly by his wife. In the Sixties there was a major rally in support of Russian Jewry at Madison Square Garden in New York. Next day a letter signed by Landau and the then much talked-about Soviet economist Yevsey Liberman appeared in *The New York Times*. It claimed that though Jewish, they both enjoyed total freedom and excellent conditions to pursue their work in the Soviet Union. At that point Landau was no longer working. This was not allowed to stand. A number of American scientists, Chandrasekhar among them, asked that the Soviet government, rather than arrange for the sending of such letters to the *Times*, allow Landau to come to the West for a visit. That was the end of it.

Unlike in the West where a physics seminar typically lasts one hour, Landau's seminars ran into many hours and consisted largely of freewheeling discussions involving all those present and often punctuated by Landau's quite cruel sarcasm. The Western physics literature was available to them, but there was no direct Western input, so the Russians were able to take a look at problems, unbiased by Western fashions and this often led them in new directions not explored in the West. It would often take years before the West paid proper attention to these remarkable new directions opened in Russia. Take superconductivity: in 1950, Landau and Ginzburg constructed a simple theory, which in the West was viewed as no more than a very clever, but far from definitive, model for this remarkable phenomenon. In 1957 Bardeen, Cooper and Schrieffer at the University of Illinois came up with their own theory based on "first principles" and superconductivity was considered understood. In 1972 they received the Nobel Prize for this theory. But in 1959 Landau's collaborator Gor'kov proved that the Bardeen-Cooper-Schrieffer theory implies the Landau-Ginzburg theory, which is therefore a different, and maybe more convenient, way of looking at superconductivity at the same deep level of "first principles." In 2003 Ginzburg was awarded the Nobel Prize for this, by then fifty-three year-old theory. Landau was dead by then, but he had received the Nobel Prize in 1962 already for his superfluidity work. Remarkably, these ideas were moved from the field of solid-state physics to particle physics by Nambu, Goldstone and others, and there it is the Landau-Ginzburg version that is favored to this day.

I could give many more examples of work in which the Russians provided entirely new avenues, emphasizing totally different aspects of a problem. The quality of the end product is invariably higher once we have more than one way to obtain it. For this to be possible, it is essential that there be two efforts — one in Russia and the other one in the West in this case — by two excellent groups that are not in strong contact. The lack of the strong contact was here provided courtesy of the dictatorship of the proletariat. This dictatorship disintegrated with the fall of the Soviet Union and the *crème de la crème* of Russian physics and mathematics moved bodily to the West. This has considerably strengthened Western physics and mathematics, while at the same time, sad to say, striking a devastating blow to Russian science. But once in the West, these Russians are no longer immune to Western fashions, and the remarkable alternative approach we so often got from Soviet science turns out to have been no more than a luxury of the cold war. Now that war is over. It is ironic that in retrospect, even a historic event as traumatic as this cold war, had some beneficial consequences.

This is not entirely limited to the realm of science. Think of music. While Western composers followed the teachings of the second Viennese school, the Russian alternative was provided by Prokofiev, Shostakovich and later Schnittke and others. Had the communist dictatorship and its hardships not occurred, would Shostakovich, like everyone else have followed in the footsteps of Schoenberg instead of those of Mahler? Poetry largely followed the same pattern, think Ossip Mandelstam, Anna Akhmatova, Boris Pasternak and others.

Physics, mathematics, music and poetry are all quite abstract and as such not amenable to rigid regulation. In fiction and in painting the end product is supposedly understandable to a much wider audience, no wonder in these fields Stalin and his commissars have wreaked much more damage.

Chandra, Passionate Cambridge Gentleman

It would be extremely unfair to judge Sir Arthur Eddington entirely by the wild and preposterous numerology of his old age, which was made fun of by Bethe, Beck, and Riezler. Sir Arthur was a great astrophysicist. He was the first to experimentally verify the bending of light by the sun as predicted by Einstein's general relativity. His work on stellar evolution is also of utmost importance. Yet, it was Sir Arthur who chased the young Subramanyan Chandrasekhar out of Europe for the very work that was to win Chandra the Nobel Prize many years later. Paradoxically, Sir Arthur was motivated to do this by a very major insight of his own which he could not accept and finally refused to believe.

Chandra was a graduate student at Cambridge in the early Thirties. His teacher R.H. Fowler had been the first physicist to bring quantum theory to bear on the problem of stellar evolution. Chandra realized that when the mass of the stars considered by Fowler becomes very large, the electrons in them can move at speeds close to that of light. He redid all of Fowler's work, taking into account not only quantum theory but also Einstein's special relativity, which applies at such speeds. Much to his surprise, Chandra found that when their mass exceeds a certain limit, — what we now call the Chandrasekhar limit, — then the stars can no longer reach the final configuration suggested by Fowler.

Eddington was interested in this work and came daily to talk to Chandra about it. Eddington immediately realized that if Chandra was right, then such stars would undergo gravitational collapse and become what have since been named black holes. Black holes have such strange properties — for instance, once an observer gets close enough to a black hole, he can no longer communicate with the world outside it, even though from his own point of view nothing out of the ordinary seems to be happening — that Eddington simply could not admit their existence. He preferred to believe that this graduate student must be wrong, although both Dirac and Fowler assured Eddington that Chandra was right. But then, this all followed from quantum theory, which itself was only six years old, so Eddington was even willing to contemplate the possibility of quantum theory itself being wrong.

Unfortunately, the way Sir Arthur chose to deal with this matter was far from cricket. Chandra was scheduled to present his theory to the Royal Society in London. Without notifying Chandra, Sir Arthur, a senior member of that august society, made arrangements with the chairman of the meeting to give his own talk right after Chandra's. He proceeded to attack Chandra virulently, almost as if he were a crank. The twenty-one year old Chandra tried to defend himself, but Sir Arthur haughtily dismissed him, "You look at it from the point of view of the star, whereas I look at it from the point of view of nature."

Afterwards things got worse. Chandra was scheduled to speak at a major international conference in Paris.

When he arrived, the organizers requested that he cancel his talk. "Sir Arthur is here" they said, "and we do not wish to create a circus atmosphere."

Chandra saw that he had no future in Europe and he wisely decided to come to the University of Chicago. As time went by, the inescapable validity of Chandra's 1931 discovery gained universal acceptance, and he ended up receiving the 1983 Nobel Prize for it. This made Chandra the second Indian physics Nobel laureate, the first having been his uncle Venkata Raman in 1930.

Years later, during the war, Chandra and Sir Arthur made their peace and Chandra behaved in a truly magnanimous way. Though Eddington had rejected the idea of a black hole, Chandra, in his talks and writings, insisted that Eddington be given credit for having been the first to have a clear picture of a black hole.

In Chicago, I met with Chandra about once every month either in his office or in mine to talk about physics, politics, and personal matters. In his formal way Chandra could be a very informal friend. It took him a good fifteen years before he first leveled with me about the Eddington episode. To imagine Chandra, careful Chandra, at the center of a controversy, was something of a shock to me, much as I by then knew that he was human.

There were two sides to Chandra. On the one hand, he was a passionate man. When he talked about physics his eyes would light up and even his carefully chosen words could not hide the fire that called them into being and the intense drive that was leading them to their destination. On the other hand, he strove diligently to approximate his ideal of an English gentleman. Chandra delighted in reminiscing about his student days in Cambridge. One of his favorite stories was about how Paul Dirac, after reading E.M. Forster's *A Passage to India*, had run into the great novelist at his college. He asked Forster, "What *really* happened at the Marabar Caves?"

"I am afraid I don't know" the celebrated author supposedly replied.

He also told about the English cosmologist E. A. Milne, who had a chair in Cambridge when Chandra had been there. One day Chandra passed Milne's office and looked in. There was Milne standing intently before the blackboard on which he had written four names: Newton, Maxwell, Einstein, Milne.

"Can you imagine this? Isn't it sad?" Chandra asked me.

For many years Chandra was the editor of *The Astrophysical Journal*, the world's premier Astrophysics publication, published by the University of Chicago Press. During the Carter Administration a woman from the Internal Revenue Service appeared on campus to question the tax-exempt status of the publication. Her argument: if the astrological weeklies one purchased at the supermarket checkout counter, all paid taxes, why should just the Astrophysical Journal be tax-exempt? The journal's staff tried to convince her that astrology and astrophysics are not the same thing, but this government official couldn't see the difference. The matter ended up in court and the University won, but not without paying out a handsome sum to attorneys. Chandra was outraged, though at the same time he got a good laugh out of this theatre of the absurd in his own backyard.

Chandra loved music. His wife Lalitha sang Indian music beautifully and Chandra loved Western music as well. He was particularly fond of Beethoven's late string quartets; it intrigued him that towards the end of his life Beethoven was able to come up with something so original. In his 1975 Ryerson lecture he discussed the difference between artistic and scientific creativity, the former running into ripe old age, the latter usually trailing off in early middle age.

A deep interest in music was shared by many great physicists and mathematicians. Max Planck and Werner Heisenberg were accomplished pianists, as is my teacher Walter Thirring, who also composes. The great American topologist Marston Morse is reputed to have been of concert pianist caliber. Edward Teller, as well as the great algebraist Irving Kaplansky and many others were fine pianists. Then there is that famous violinist Albert Einstein. John Eaton, the opera composer, once told me that as a composition student at Princeton he was sent over by the Music Department to Einstein's home to accompany the fiddling physicist on the piano. He did as told, though not without apprehension. He expected a total dilettante. Remarkably enough, he found Einstein's musical instincts astute and on the mark. He really understood how one had to play the piece, and had he but commanded a better technique, he would have been a truly fine violinist.

I can't think of many great writers (Jean-Jacques Rousseau?), painters (Paul Klee, but forget Ingres, truly a dilettante), poets, sculptors who were also accomplished musicians. Why is it precisely mathematicians and theoretical physicists who surrender to Euterpe's spell? It is often said that music shares with these sciences an aura of rich abstraction. Music even has number-theoretic underpinnings going back to the ancient Greek mathematician Pythagoras. Though all this may indeed play a role, I think there is another reason as well. Mathematicians and theoretical physicists deal with extremely beautiful abstract ideas. The creative process, even in these sciences is a highly emotional, indeed passionate activity. Yet these emotions and passions are of necessity suppressed in the end product, the scientific paper, which is presented as the result of careful cold reasoning. There is left at the end of the day an unfulfilled yearning to express these deeply felt emotions, and music offers the ideal venue for doing this, whether actively by actually performing the music, or passively by listening to music and letting one's heart roam freely in its fields.

Chandra completely changed field about every seven years. He didn't just move on to a new problem, but embarked on a new area of physics with the passion of a young lover. When I arrived in Chicago, he was just moving on to general relativity. The hero of this field is Einstein, and Chandra, not one to shy away from a heavy dose of hero-worship, became a great Einstein-worshiper. He would come to my office and for long stretches hold forth about Einstein's supreme genius.

But all this changed when in 1987, for the tercentenary of Newton's *Principia*, Chandra was invited to write about this unsurpassed volume of fundamental scientific discovery. He told me he didn't feel like reading the whole book. Instead, he decided that he would read Newton's own wording of the most important theorems proved in it. Then he would sit down, armed with Newton's calculus, and derive all these results, the way a good undergraduate student would do nowadays, and compare his proofs with those of the master. But much to his surprise, none of his proofs bore the slightest resemblance to Newton's. Intrigued, Chandra started reading the proofs given in the *Principia*. Although Newton had invented calculus, he used it only sparingly in the book. Where a routine calculus-based proof could be avoided by an extremely clever use of elementary geometry, Newton didn't hesitate to go that route. Chandra was awed. Suddenly Newton became his new hero.

Chandra decreed that these geometric proofs were *the* evidence of Sir Isaac's unsurpassed genius. There is no doubt that as a scientist Newton had no peer; he was very likely *the* greatest scientist who ever walked this earth. Yet, regarding the *Principia* proofs, one has to allow for the possibility that their geometric nature was dictated not only by the needs of genius, but also by some more mundane factors. For one thing, at that time calculus was still in its infancy, and many potential readers of the *Principia* could not have been expected to have mastered it yet. The geometric proofs went a long way towards making the book accessible to those readers. Also, Newton himself was brought up in the tradition of the "old" mathematics, with its emphasis on elementary geometry, and whenever possible, people tend to do things the way they were taught to do them. To this day my mother calls on her old rotary phone and finds no use for push-button tone-dialing, it's what comes naturally to her. Could it be that at this exalted level, elementary geometry was Sir Isaac's rotary phone?

When I raised these questions with Chandra, he remained adamant: it's all just Newton's genius, period. He then upped the ante. With Newton freshly promoted to superhero status, there was no place left for Einstein. Suddenly Chandra began finding immense faults in Einstein. To discover general relativity, and then to calculate the gravitational field of the sun only approximately when an exact solution could have been found and was found by Schwarzschild? Simply unforgivable. True, the approximate solution worked out by Einstein was sufficient for comparison with astronomical observations — still can you imagine Newton having missed the exact solution? Einstein was neither a mathematician nor an experimental physicist in the same league with Newton, but their contributions to theoretical physics *are* comparable.

Chandra really turned on Einstein and it fell on me to defend him, not that the old boy couldn't fend for himself. I loved these discussions. They were intense, passionate, and logical, and they did not trespass on the territory of our friendship.

Occasionally Chandra and I disagreed on political matters, one in particular which temporarily strained our relationship. It had to do with the 1972 presidential election. Chandra was wearing a big McGovern button on his lapel. I couldn't stand the man because he sounded so self-righteous. By contrast, I found the Nixon-Kissinger foreign policy imaginative, much as I disliked Nixon's domestic policies. On the assumption that Nixon's main motivation was to leave the legacy of a great president, I suggested that in his second term, when he didn't have to continuously worry about his reelection, he would bring in a "domestic Kissinger" as well. But Chandra, like most of my colleagues found this argument preposterous — my popularity took a big dive. Chandra showed up at my office in the summer before the election. He told me agitatedly that he had been in Munich and had visited Heisenberg.

"We talked about Nixon, and Peter, you will be very pleased to know that Heisenberg fully shares your views on this matter." He delivered this ultimate putdown with oozing sarcasm. So be it, I thought at the time.

Nixon was reelected, and then came a much-publicized interview Kissinger gave to the brilliant journalist Oriana Fallaci, the one in which he compared himself to the lone cowboy riding away into the sunset — implying that ending the Vietnam war was his own doing. I immediately started worrying that, to prove that he deserved most of the credit, Nixon might do something wild. Sure enough — there came the Christmas bombing of Hanoi. Rightly or wrongly, I took *that* to be Nixon's response to Kissinger's Fallaci interview. Then I ran into Chandra in the Fermi Institute library.

"I have to admit you were right on Nixon. This bombing is outrageous."

A big smile settled on Chandra's face. "I am just as outraged as you are, and have already drafted a letter to the U.S. Senate urging them to censure the president. If you *really* mean what you say, then you should cosign this letter with me."

Our letter went out that same morning. The Senate did not censure anybody, but, as it turned out, these same senators had Watergate up their sleeves. By the time *that* show made it to prime time TV, I could watch it in the privacy and comfort of my own living room with a clean conscience, well, maybe not as clean a conscience as Chandra's.

When Chandra retired, I asked him whether he intended to return to his native India. No, it would make no sense, he replied without hesitation. Sensing my surprise at his certainty, he brought up the name of D.S. Kothari, his distinguished Cambridge classmate, who early in his career had returned to India and had played a decisive role in building a first rate school of physics there. Had Kothari stayed abroad, he may have achieved more as a scientist, but he knew he had to go back. Chandra on the other hand, found himself neither interested nor gifted at science administration and organization. Like me, on this matter he was a Marxist — more precisely a Groucho Marxist, for neither of us would ever want to belong to a scientific organization that would have either of us as its administrator. His only meaningful alternative was therefore to stay in Chicago, doing what he did best: research that is. But he added philosophically, "In the long run, maybe Kothari's contribution will turn out to have been the more important one by far."

23

On and Off the Map: Romanian Mathematics

To fully appreciate Kothari's achievement on behalf of Indian science, consider the story of what has happened to science, and more specifically to mathematics in Romania, another latecomer to modern scientific prominence. The Romanian case highlights the perils that lie along the road towards a viable science enterprise.

Like India, Romania was also divided into principalities and underwent colonial rule. But whereas in India the colonial power was Great Britain and the colonial status lasted for only about one century, in Romania the colonial power was Turkey and colonial rule lasted for many centuries. Like Greece and other Balkan lands, Romania acquired its independence in the second half of the nineteenth century, just about the time when Queen Victoria became Empress of India. At that point a true cultural revolution in the best sense took place in Romania. Literature flourished (above all the playwright Ion Luca Caragiale, the great forerunner of the theatre of the absurd), the arts soon followed (Constantin Brâncuşi, Victor Brauner, Tristan Tzara) as did music (George Enescu, Dinu Lipatti). Young

Romanians were being educated in the West, especially in France and Italy. So by the turn of the nineteenth century, the first generation of significant Romanian mathematicians appeared on the scene, Dimitrie Pompeiu, Gheorghe Tzitzeica, Traian Lalescu, and others. These men stood tall as role models for the next generation, which would include four internationally recognized figures: the renowned function theorist Simion Stoïlow (much cited by the American Marston Morse), that jack-of-all-trades Grigore C. Moisil, the geometer Gheorghe Vrânceanu and the analyst Miron Nicolescu. Romanian mathematics has been, so to speak, put on the map and was developing fast.

But history can be counted on for serving up some unwelcome surprises. Whereas once Indian independence was achieved, it held, the same cannot be said of Romania, for after less than a century of independence Romania became a Soviet colony, or satellite state, to use that Western euphemism. For a while, even under the Communists, mathematics kept rolling along, although certain fields — those labeled "idealistic" in Moscow — became taboo. Among these fields were mathematical logic and the study of the foundations of mathematics, in general and, in particular the modern developments in the foundations of geometry. One of my teachers, Iosif Kaufman, found himself in the unenviable position of having to teach foundations of geometry as if in the two millennia since Euclid nothing had been done in this field. In fact major progress had been made by David Hilbert in Germany, but, in view of Moscow, Hilbert was a proscribed idealist. So what was Kaufman to do? He could certainly pretend Euclid's was still the last word in the field. He could even go as far as suggesting that at heart Euclid had been a pre-Marxist Marxist and we are talking Karl, not Groucho here. No one would have contradicted him and risked deportation, just to set the Greek record straight. What Kaufman did was much more original.

After lauding Euclid's solid philosophical credentials, he went on to warn his students that the idealists and imperialists, in their diabolical attempts at undermining Marxism-Leninism-Stalinism, have enlisted the clever German mathematician Hilbert's services, and that against a fat remuneration from his imperialist masters, Hilbert had concocted a vile mathematical brew perverting Euclid's Marxist achievements. So insidious and devilish was this work of Hilbert, that young Romanian mathematics students had to be exposed to its perfidy, in order to develop the vigilance befitting future communists. He then went on to lecture on Hilbert's gorgeous work, and in the last lecture he reminded his students that now they could consider themselves fully immunized against it.

Grigore C. Moisil, who received me at his home each time I went to Bucharest and acted as a valuable sporadic teacher, once gave me two proscribed books which he had bought in Ankara some time earlier. Both books dealt with mathematical logic: one was by that same David Hilbert, and a collaborator, whereas the other, titled *A System of Logistic,* was by none other than Harvard's Willard Van Orman Quine, the American devil incarnate himself, a man who did not worship at Karl Marx's altar and yet dared call himself a philosopher. Moisil warned me "Hide these books well, but if through carelessness or stupidity they find them on you, tell them what you want about how you got them. But under no circumstances let it be known that I gave them to you. Do I have your word?" I gave my word and I made good use of the books.

I left Romania in 1959, before the tragic Ceauşescu era. The troubles of Romanian mathematics started before Ceauşescu and culminated under that tyrant's rule. In 1961 two of the most talented mathematicians of the next generation obtained their doctorates, and invited their teachers to a celebratory dinner at a Bucharest restaurant. Moisil and

Stoïlow were both at the dinner. These two did not get along well. Moisil did not hide his disdain for the regime nearly as well as well as he did his books on mathematical logic. Stoïlow, by contrast was a communist and had been one even before that became fashionable. He was also a thoroughly decent human being who could be counted on when one of his students got in trouble. Being both a communist and the more distinguished of the two, Stoïlow became the chairman of the mathematics section of the Romanian Academy of Sciences. Aside from their political differences, Mosil and Stoïlow were two big fish in a small pond and they did not see eye to eye at a personal level either. There was one more difference: Moisil was quite partial to the bottle, so at this dinner before long Moisil was loudly mouthing the politically incorrect. Stoïlow made it clear that if Moisil went on spouting his heresies, Stoïlow would have to denounce him, or else he would get himself in deep trouble for having idly stood by, when news of these goings-on were reported to the Party, as they were sure to be, given the dinner's large attendance. Moisil, quite inebriated by then, wouldn't listen, so Stoïlow got up and left. Next day, just as he had warned Moisil, Stoïlow was on his way to the Central Committee to denounce his colleague. This did not come easy to a thoroughly ethical man, and on the very staircase leading to the Central Committee Stoïlow was felled by a stroke.

Under such circumstances, Moisil obviously could not be elected to replace Stoïlow as the chairman of the mathematics section, so the academicians settled on Miron Nicolescu, an affable, debonair man, and a very handsome ladies' man. Nicolescu once came to give a talk at the Technical University in Timişoara where I was a student. Given his fame, the talk was scheduled in the large hall of the department of mechanics. Usually, in the afternoons this hall was empty and the students used it as a kind of lounge. Not many people attended Nicolescu's talk, and every other minute or so a student coming to lounge would open the door to the hall, take a surprised look at the speaker, squint at his formulas and then leave, more or less slamming the big door. When Nicolescu was finished, a lady who traveled with him was also called upon to speak by the chairman of the seminar, evidently as a condition of Nicolescu's appearance — "Either she speaks or neither of us speaks." The lady turned out to be an astronomer. As it happened she was also very young and beautiful, and fashionably, if somewhat gaudily dressed; she had a good ten bracelets, a necklace, rings on her long fingers, and her nails polished in crimson. In a day and age when female astronomers were very, very few in Romania, — the sciences in general were a male preserve — her appearance created quite a stir. The would-be loungers now opening the door did not retreat in horror as during the academician's talk, but stayed in the lecture hall. The few that still elected to leave did so only to return soon after with friends wanting to witness this eighth wonder of the world. By the time the academician's companion had discussed all the stars she had been watching, the hall was full. Her talk was greeted with immense applause and loud cheers.

Nicolescu, now chairman of the mathematics section of the Romanian Academy of Sciences, soon became president of the whole Academy. Fast-forward to the Ceauşescu era, which started in 1966. Of the "Great Leader's" three children, Zoe was a mathematician and his son Valentin a physicist. One evening in the mid-seventies Zoe, by then a research scientist at the mathematics Institute of the Academy, did not come home when she was expected for dinner. Like any loving parents, the comrades Ceauşescu, grew more worried with each passing hour that their child failed to come home. After waiting for a while like this, the "Great Leader" called the Securitate, his own secret police force, and ordered them to find his beloved daughter. In a country where even shadows were shadowing shadows, Zoe was quickly located not far from Bucharest, in a motel in the posh mountain resort of Predeal, where she had checked in for the night with one of her colleagues from work. While Zoe was returned unharmed to her fretting parents, the colleague's trail seems to end here traceless, in the best mathematical sense of that word.

What would you do, if you had your own secret police force and had command over your academy, and your daughter went to a motel with a man working in the Academy's mathematics Institute rather than come home for dinner one evening? I don't know what I would do, but I can tell you what the Great Leader decided to do. He abolished the mathematics Institute of the Academy, just like that. When he learned of this, Miron Nicolescu went to Ceauşescu to plead the case of mathematics. He tried to explain that it took almost a century to build up a viable mathematics tradition in the country, and that this all could be undone with one stroke of the pen. "My wife tells me, if *that* is what they do at the Institute of mathematics, then Romania needs no mathematics" was the Great Leader's memorable reply.

And his wife should have known. She was a member of the Academy herself. When her husband made it to the top, not surprisingly she wanted a doctorate even though she had not even graduated high school, if she ever attended one. She decided that chemistry would be her field. She went first to Constantin Neniţescu, the country's leading chemist and when he refused to grant her the instant doctorate she sought, she had him forcibly retired and found herself a thesis sponsor, a thesis writer, and whatever else it takes to get the coveted degree. Once she was Dr. Elena Ceauşescu, she got herself elected to the Academy and kept up a steady outpour of papers crafted by her assistants and bearing her name as senior author. I could not ascertain that she ever read them or even that had she actually been able to read any of these papers, she would have chosen to do so.

In view of the good relations between Ceaușescu's Romania and Iraq, the Iraqi Academy of Sciences, probably at the urging of Academician Saddam Hussein, published the collected works of Dr. Elena Ceaușescu. So, strictly speaking, the abolition of mathematics in Romania was not the capricious act of a moronic dictator, but rather the well thought-out action of a literate chemist of world renown.

Abolishing the mathematics Institute, even in a country that "needs no mathematics," is a devastating business. The members of this Institute who held professorships at various universities could return to their teaching with the corresponding cut in income, but those without such positions were sent to villages to teach elementary school. Shortly after this, Miron Nicolescu died of a heart attack; I guess his heart literally broke. The chairmanship of the mathematics section of the Romanian Academy was apparently hazardous to that position holder's health. Of course I have no good reason to rule out the possibility of a curse.

For a few years the Romanian mathematics journals stopped coming to Eckhart library at the University of Chicago. Then suddenly they started arriving again, nicely refurbished with an impressive silver cover. The new editor was Zoe Ceaușescu. A whole generation of Romanian mathematicians had been eliminated by the family. Ciprian Foiaș, the famous Romanian mathematician, once aptly described the Ceaușescus to me as *lumpenproletars*.

The Ceaușescus eventually got what they asked for. They were hunted down like dogs by armed soldiers at the end of a trial carried out in a manner they would certainly have approved of for anyone but themselves. Remarkably, in their heyday this nefarious couple had the support of a number of Western statesmen. Charles de Gaulle was visiting the Ceauşescus when the youth rebellion started in Paris in 1968. He rushed home but it was too late. Richard Nixon also put in an appearance in Bucharest, he even quoted the Romanian poet Mihai Eminescu (one wonders whether he had the slightest idea who Eminescu was), but like de Gaulle before him, after shaking hands with the Ceauşescus, his fate was sealed. Still skeptical about curses? At least de Gaulle and Nixon got away with their lives.

This story was common knowledge in the American mathematics community, but to the best of my knowledge has never appeared in the American press. It's a good story, a tyrant, illicit sex, you name it. Why was it not deemed "fit to print?"

On the sunny side, after the 1989 revolution, Romanian students were again flocking to Western universities. Some are very talented. Some stay in the West, but some will return and put Romania back on the cultural map, where it belongs. Good Romanian science will go on, but it is an ephemeral thing at the mercy of the neighborhood's next *lumpenproletar*.

24

Du côté de chez Telegdi, the Experimental Side

I have wandered in the world of the theoretical physicist and of one of his partners, the mathematician, somewhat to the exclusion of the other partner, the experimental physicist. Remember he is the one who lives on the lowest of the three planes, the only plane that touches the ground and is firmly grounded in reality. To introduce you to this experimental world let me start with the one experimentalist who was closest to me, or at least to whom I felt closest. A remarkable, thrilling and exciting man, the Hungarian-born Valentine Telegdi made fundamental discoveries in physics. Valentine with Jerry Friedman performed in Chicago one of the three key experiments to confirm the Lee-Yang proposal that nature knows how to tell left from right. Telegdi nailed down quantitatively the Feynman-Gell-Mann-Marshak-Sudarshan proposal of the detailed way in which nature goes about telling left from right. Forty years ago at CERN, with Dick Garwin he led the first of a crucial, still ongoing, series of "muon g minus 2" experiments, which established that the mysterious muon is for all practical purposes a bloated cousin of the electron, while at the same time testing with high precision the Feynman-Tomonaga-Schwinger renormalized quantum electrodynamics. These are just three of his brilliant experiments. What makes these experiments so remarkable is their subtlety.

Take the "muon g minus 2" experiment. In a famous book Sir Nevill Mott, the British Nobel Prize winner, claimed that such experiments are impossible for deep theoretical reasons. Most experimental physicists are masters at electronics and data analysis. They know enough theory to fully understand what they do and why it is important, but they leave theoretical subtleties to the experts. Valentine was the rare experimental physicist, who had such a mastery of things theoretical, that he could form his own ideas about what should be measured, what can be measured, and how it should be measured, with uncanny elegance and economy of means. Valentine took on Sir Nevill. Some earlier electron experiments and calculations in special cases had shown that Mott was wrong. Yet, it was Valentine's classical theoretical paper with Valya Bargmann and Louis Michel — two very theoretical theoreticians — that conceptually clarified the issues in their full generality. This paved the way for the high precision muon experiment.

Another theoretically very savvy experimental physicist is Maurice Goldhaber, who settled the question of whether neutrinos are left- or right-handed by using a particular sequence of nuclear decays, so cleverly chosen, that no other sequence could have been used. Telegdi and Goldhaber shared a Wolf Prize, second in prestige only to the Nobel Prize.

Valentine was always interested in hearing the latest theoretical developments, in character for a man who did experiments the way Valentine did. He would reciprocate. In the good old Sixties the theoretical and experimental planes were very close. When a new experiment came out, Valentine would give me his expert opinion as to its strengths and weaknesses. An experiment performed at Brookhaven National Laboratory's proton accelerator in 1963 by Sam Lindenbaum and Luke Yuan confirmed the relevance of the ideas of the Italian theoretician Tullio Regge, and in the great effort to further develop these ideas string theory was born. But Valentine was fascinated by the way this experiment had been carried out.

"This experiment has revolutionized the way we do experimental physics. No experiment will ever be done again without an online computer" he told me.

Before Lindenbaum-Yuan, experimentalists working at particle accelerators would place their equipment in the path of the beam, leave it there for as much time as they had been allotted and then remove it and analyze the results. This analysis could hold some major bad surprises. From time to time one of the graduate students, research associates, or even senior physicists involved in the experiment would forget to push that green button or place that counter where it was meant to be — the whole experiment was therefore in vain. You may have to wait five years for more time; there is only one accelerator and many users are lined up to do their experiments on it. On the less fatal side, you could find three really tantalizing events in the data. Had you been able to monitor what you were doing, you could have adjusted the equipment so it picks up more events of this type. By placing a computer online, Lindenbaum-Yuan monitored the experiment as it was running. Any troubles? They could correct the problem before it was too late. Like many great ideas, this one sounds trivial yet represents a real breakthrough. It was 1963, and their state-of-the-art computer would be rejected now by a six-year old as too primitive. But what a difference it made then!

Luke Yuan, Lindenbaum's partner in this experiment, came from a prominent Chinese family. He was the grandson of General Yuan Shih-k'ai, who ruled China right after the fall of the Qing Dynasty in 1912. This fact was used maliciously during the Cold War. After the Sino-Soviet split in the Sixties, the Soviet scientific establishment tried to mend fences with the West. They invited an American delegation and insisted that Luke Yuan be part of it, just to infuriate the Chinese. Luke Yuan was married to C.S. Wu, one of the greatest woman experimental physicists of all time. Of the three experiments that proved that nature can tell left from right, the experiment led by C.S. Wu of Columbia University was the first one to be completed, followed shortly by those of Garwin, Lederman and Weinrich also of Columbia and of Friedman and Telegdi at the University of Chicago. This was something that had never happened before. Three spectacular, independent experiments proved the same fundamental property of the laws of nature, within days of each other. Since that time we have witnessed some more such coincidences, for instance, the discovery of particles containing the "charmed quark." This is the kind of thing that happens when the number of physicists exceeds a certain threshold and the speed of communication is very high. Physics is no longer the leisurely enterprise it had been once upon a time. This intense competition wears the nerves of those involved and occasionally conflict can ensue. It would be dishonest to pretend such things don't happen, they do, physicists aren't these calm elderly gentlemen, wearing white lab coats and thick horn-rimmed glasses. Only in Hollywood can you still see these stereotypes. On the contrary, physicists are, as should be clear by now, highly emotional creatures, fueled by dreams and by ambition and all this works out for the best, as if Adam Smith's invisible hand itself were waving over our heads.

Maybe the most famous of these conflicts is the one between Valentine Telegdi and C.S. Wu. As if one such collision were not enough for a lifetime, the two of them managed to cross swords on another series of experiments as well. During the controversy Telegdi and Wu communicated through vitriolic letters most rigorously observing all the niceties of both Hungarian and Chinese etiquette.

Once during a conference at Fermilab, the big laboratory near Chicago, this hostility spilled out into the open. Telegdi was scheduled to speak right after Wu. They each were allotted half an hour. Years earlier, Wu had done a famous experiment in which she discovered a weak interaction counterpart to ordinary magnetism, "weak magnetism," a fundamental feature of the standard model of particle physics. But a few months before the Fermilab talk, it was pointed out that certain nuclear physics tables she had used were incorrect. On the face of it, this meant that the size she had claimed for the weak magnetic effect, would have to be changed, leading to the collapse of the standard model. Her talk was intended to convince the audience that upon reexamining her old data, she had found a further mistake along with the use of the wrong tables, and this mistake precisely canceled the mistake due to the tables, leaving her original final result unchanged. She took a long time to go through all this and Abdus Salam, who chaired the meeting, approached her repeatedly at the lectern to tell her that time was up. Each time she politely smiled at Salam and just went on with her lecture as if nothing had happened.

Finally, after taking up fifty minutes, leaving barely ten minutes for Valentine's talk, she came to the end. At this point Salam called on Telegdi, who went to the podium and said, "I have prepared a talk which I cannot possibly present in the ten minutes left me by the previous speaker, so I will simply not give my talk. I should add though that on the whole a good purpose will have been served, for as the previous speaker has made amply clear, the magnetism she found is truly *weak!*"

Michael Atiyah, the mathematician, was sitting next to me. "What is going on here?" he asked me. I gave him a brief rundown on this famed feud.

"You have these things in physics too? I thought this only happens in mathematics." Just to avoid any possible confusion, the amount of weak magnetism predicted by the standard model of particle physics is experimentally well confirmed by now. I should add that in person Ms. Wu was a charming colleague with whom it was a pleasure to discuss physics, and what physics at that. But she was also a very passionate and forceful woman and, God, could she write those letters to Valentine.

From Luke Mo, one of her former students, I learned that in her laboratory at Columbia she was also a strict disciplinarian. When Luke Mo's first child, a son at that, was born, Luke wanted to rush to the hospital to see his newborn. C.S. Wu told him in no uncertain terms, that were he to leave the lab early, he need not come back. He stayed in the lab, he told me, with tears streaming down his cheeks, years after the event. But science demands total devotion and does not tolerate any interruptions, no matter how compelling the reason. C.S. Wu obviously taught this important lesson to her students in her own way. In her defense, let me add that it is said that she was the daughter of the last Qing emperor's fifth wife, and this may explain her strict ways.

This strict discipline enforced by C. S. Wu in her lab was the focus of another of her confrontations with Telegdi. At a Caltech seminar she spoke about her work on double beta decay, which she performed underground in a deep salt mine to minimize the cosmic ray background, Telegdi quipped "this confirms what I always suspected — it is better to work in a salt mine than at Columbia University."

Valentine Telegdi, who died last year, was a charming and above all thoroughly frank and honest person and the ultimate connoisseur in culinary matters. He was the first person to whom I confided that I was going to marry Lucy, whom he had met by then on a number of occasions. He looked at me horrified

"That woman? You don't know?"

"I don't know what?" I worriedly asked.

"She is American."

"So what?"

"What are you going to eat?"

If anyone had the right to ask this odd question, it was Valentine, whose Italian wife Lia, is by far the best cook I have ever met outside a three-star restaurant. As it happens, my wife is not that far behind either and each time Valentine came over for dinner, he repented for his remark, as if our dining room had suddenly turned to a confessional booth.

After attending a conference in Europe, I brought back the freshly released original French edition of Paul Bocuse's cookbook. In it he gives the recipe of the dishes he had invented for Valéry Giscard d'Estaing's presidential inauguration dinner. Lia and Lucy got together and cooked the whole shebang, truffle soup under crust and all, for us. It was the best meal I ever had in a private home. Many physicists are gourmets. Murray Gell-Mann was a frequent diner at the Telegdi home. Given Valentine's penchant not only for food, but for its almost homonymic feud, Lia asked Murray

"At any given time Val has to have a person to hate. Is this really necessary?"

"Look Lia" Murray answered, "somehow Valentine in many cases picks the right people to hate."

In the Seventies, Valentine Telegdi accepted a professorship in Zurich and our Chicago faculty meetings lost much of their drama, though staid Zurich became a much livelier place, I understand. This is not even to speak of the immense contribution made by Valentine to the scientific atmosphere of the institution he called home.

Still, Chicago's record in experimental physics stands as one of its major glories, starting all the way back with Albert Michelson. Leon Lederman, another one of the Telegdi's competitors on that legendary trio of near-simultaneous experiments, also became a member of our faculty. Belatedly in 1988 he did share a Nobel Prize with Mel Schwartz and Jack Steinberger for discovering that the muon's neutrino is a different particle from the electron's neutrino, although he did quite a number of other Prize-worthy experiments as well. On one occasion before 1988 his son asked him why he hadn't yet received the Prize. Leon is said to have answered "I deserve it for so many experiments, the Nobel committee can't make up its mind which one to give it to me for."

Finally, Jim Cronin put his stamp on Chicago experimental physics. With Val Fitch he made an epochal discovery. After it became known that nature can tell right from left, people started asking the question "What does an electron see when it looks in a mirror?" for mirrors, as we all know, exchange left and right. When he completed his experiment confirming the Lee-Yang proposal that nature can tell left from right, Telegdi ran into Leo Szilárd on campus. Though by then heavily into molecular biology, Szilárd, a brilliant physicist, kept up his curiosity about new physics developments. As soon as Telegdi told him what he had found, Szilárd asked the very question I have just mentioned. But he went on to answer his own question: "Maybe

in the mirror the electron sees its antiparticle, the positron."
A number of people came up with the same idea and fleshed
it out. Experimentally it looked as if they were right. But just
at the moment when people started taking this idea for
granted, the Cronin-Fitch discovery shocked the world. Even
that is not true! It is not sufficient to exchange left with right,
matter with antimatter; you have to also exchange the past
with the future. In a sense, it is this Cronin-Fitch discovery
that to a large extent is responsible for our existence. We are
made of matter, and a friendly habitat better be free of antimatter, or else we'd get annihilated just like that. There must
be more matter than antimatter in the universe for this to be
possible and it turns out that if matter saw antimatter in the
mirror without the additional exchange of the past with the
future, no such matter excess could develop. It was a heroic
experiment, this Cronin-Fitch work and it had immense
implications for cosmology.

The nature of experiments in particle physics has radically changed over the last thirty years. Gone are the days
when a Friedman and a Telegdi can make an epochal experiment. Same for a Cronin, a Fitch, a Christenson and a
Turlay, or a Wu, Ambler, Hayward, Hoppes and Hudson.
Nowadays a particle physics experiment costs in the hundreds of millions of dollars, not including the billions for
building and running the accelerator. The typical experiment
is performed by hundreds of collaborators from all around the
world. What will happen to the field when the next generation
of machines runs its course? How far will governments be
willing to go to put up money for this endeavor? The new
trend uses the universe itself as a cheap accelerator —
extremely high energy particles can be found in the cosmic
radiation which bombards us all the time. Admittedly this
cosmic radiation is not as well under our control as a beam
at a man-made machine, but then that's the kind of handi-
cap human ingenuity and genius are meant to overcome.

CHAPTER

Leader of the Pack

From a certain distance, the world of theoretical physics reveals a clear pattern. At any given moment there is an easily identifiable leader of this world. He does not impose himself, but is anointed, as it were, by the community. This leader sounds one of the main themes, and then he gets to set the style. Not that physicists look for guidance — most are highly individualistic, and some may contribute work as important, or maybe even more important than the leader. In the end it is the leader's word, his seal of approval that somehow creates consensus.

Most colleagues agree with me that one can trace leaders at least as far back as the end of the nineteenth century. Then the leader was the Austrian Ludwig Boltzmann, who with his statistical mechanics and understanding of entropy rang in the era of modern physics. Boltzmann had a long history of manic-depressive illness. In 1905, while a guest at the Prince of Torre e Tasso's castle at Duino, the same castle where Rilke wrote his elegies, he hanged himself. His era came to an abrupt end. That very year a twenty-six year-old Albert Einstein wrote five papers that revolutionized physics; a new leader was at hand. The Einstein era lasted to about 1925 when, as we saw, Heisenberg appeared on the stage. The Heisenberg era endured till 1943 and was followed by a transitional period dominated in a sense by Enrico Fermi. During this Fermi era physicists began exploring the subnuclear realm using ever more powerful accelerators. Then in the early Fifties Murray Gell-Mann was anointed and the Gell-Mann era extended into the early Seventies, followed by the Gerard 't Hooft era and finally beginning in the Eighties the Edward Witten era whose end is now approaching.

Of course there is much more to physics than these eras. it is just so that Gell-Mann and not Feynman was the leader whose approval was so dearly coveted by everyone, and who ultimately set the tone.

What is it that makes leaders of certain brilliant individuals and not of others who are equally brilliant? The names of Dirac and Feynman, to name but two, have not shown up. Are their contributions any less worthy than those of the leaders of the eras in which they lived? Hardly. Charisma plays a role. Boltzmann, Einstein, Heisenberg, Gell-Mann all are charismatic figures. By contrast, Dirac would have frowned on the very concept of charisma, though Feynman was probably the most charismatic scientist ever, the Bill Clinton of physics. Then, 't Hooft and Witten would rate lower on the general public's charisma scale. Yet to a physicist, they too conjure up a manifest charismatic aura, by their very dreaminess and by the way they seem to really feel where we are headed. They have an ability to feel, if not to outright see, the future.

Leaders share an ability to sense when something truly fundamental is coming our way, to announce to the community that "This is it" and to be right on the money. Boltzmann knew atoms and molecules were here to stay: he understood the importance of Loschmid's calculation of the diameter of a molecule. Einstein, was the first to recognize the immense importance of Prince Louis de Broglie's electron waves. Without Einstein's endorsement of de Broglie's work, the Schrödinger equation might have been found quite a bit later. Einstein was also the first to understand the importance of the work of the then-obscure Indian physicist Bose on quantum statistics. Similar examples can be given for all the later leaders as well.

To fill this function, the leader has to live deeply rooted in the physics community and that's where Dirac and Feynman missed the mark. At heart they were loners, each living in his own marvelous world. Neither Dirac nor Feynman were avid readers of the current physics literature. Both operated on the principle, "If I'll need it I'll just figure it out for myself. Why waste my time reading everything that's fashionable this month?" I remember once after the beautiful work of Feynman and of Ludwig Faddeev and V.N. Popov had been published, I told Dirac that now we know how to quantize gravity, to provide it with a microscopic description, but the trouble is, it isn't renormalizable (the techniques invented earlier by Feynman and others to deal with the unphysical infinite results obtained in calculations, do not apply for this theory). Dirac didn't hear the second part of my sentence, and as to the first part he just replied "You mean a theory containing infinities like quantum electrodynamics?" It was clear that the renormalizability of quantum electrodynamics was something he hadn't bothered to understand because as far as he was concerned, this was not the way to go, period.

As for Feynman, at one point in the Fifties he left physics altogether and decided to do molecular biology. "I have traveled all over the world" he told me, "and after a time I lost interest in going to yet another exotic corner of the world. I decided I'd rather go to another world and I started working on biology." He stayed in biology for a year or so. Had he stayed a little longer, he might have made a major discovery, for I am told he was on the right track.

But he returned to physics instead "I made my comeback with the V-A theory" was the dramatic way he put it to me. A true leader cannot leave the field: that would be tantamount to an abdication.

One final ingredient goes into making the leader: he has to be viewed by the whole community as fair, decent, and honest to a fault. I can think of physicists who don't meet this criterion, even as they meet all the previous ones.

A similar leadership structure can also be distinguished in other sciences and arts as well. Consider the history of opera. After its Italian founding fathers: Vincenzo Galilei (Galileo's father), Monteverdi, and Cavalli, opera went baroque under the leadership of Rameau, Lully and Haendel, and then revealed its full potentialities in the era of Mozart. This model was then perfected and carried further, in the direction of romantic grand opera, under the leadership of Rossini. His era was followed by that of ... Meyerbeer, whose fame exceeded by far that of any of his competitors, and whose *Robert le Diable* marks the moment in which grand opera and the romantic ballet truly arrived, were here to stay. On Meyerbeer's death, opera split into an Italian and a German branch, with Verdi and Wagner as the respective leaders. Next era: Puccini on the Italian side, Richard Strauss on the German. This is not to belittle the Russian, the French and other operas, but I am talking of the perceived worldwide leaders. Then while the twentieth century marches on with some great operas still being composed by Bartók, Berg, Janaček, Poulenc, Ligeti, Messiaen, there is no longer an obvious leader. It is fair to say that the age of the "opera composer" as such has come to an end, and opera itself as an art form may have reached the end of its glorious road. In Otello's words: "*Ecco la fine del mio camin. O Gloria....*"

But in physics we are happilly still in a period of ascent, we still *have* a leader and one hopes major new discoveries in store. It is interesting how each of these eras in physics has its own central issue and its own style.

The main achievement of the Boltzmann era was the realization that the widely explored laws governing heat phenomena, the so-called laws of thermodynamics, are not basic laws themselves, but can be derived from the then-conjectural atomic and molecular substructure of matter. The Einstein era set as its task the discovery of the fundamental laws which replace the "old" physics in the exploration of nature at microscopic distances and at large speeds. The Heisenberg era then synthesized these laws into a general theory. The Fermi era initiated a period of close collaboration between theory and experiment. During the Gell-Mann era new material was brought from experiment into the theoreticians' plane so that the exact form of the fundamental laws could be nailed down. During the 't Hooft era, this nailing down of the new fundamental laws took place with so much success that all the experimentally acquired knowledge was accounted for and theoreticians moved away from their experimental friends, not without mutual recriminations. The Witten era now confronts the tasks of including gravity in the theory, and of unifying our ideas. In the absence of any hints from experiment, mathematics becomes the conceptual basis on which we build. In Nambu's words "physics has gone post-modern."

26

New Dimensions

The four concepts in terms of which physics is phrased to this day have been handed down to us by the founding fathers, Galileo and Newton. They understood that physics is about *matter* moving in *time* through *space* under the influence of *forces.* Here are all our four concepts: *space, time, force,* and *matter.* A lot has been done in the past to reduce the number of basic forces and of basic forms of matter through unification, as I have already mentioned. The great insight of the past three decades is that all four basic concepts are really only different aspects of one and the same "agency," the *string.*

One of the crucial developments that led to this sweeping statement was the, now almost a century old, realization that we cannot take for granted even something as "intuitively obvious" as the three-dimensionality of space as seen by the naked eye. Maybe there are more dimensions that for some reason cannot be seen by the naked eye. Yet, even if the number of dimensions turned out in the end to be three, as our eyes seem to tell us, it still remains our duty to account theoretically for this fact. Why is it three and not two or 10,379 for that matter?

A good starting point is to see what would happen if it was *not* three. This question was first raised by the brilliant Finnish physicist Gunnar Nordström as early as 1914. But these days he gets hardly any credit for his bold idea. It is a rather sad story, but with a glimmer of a happy ending. Like others before and after him, Nordström was trying to unify gravity and electromagnetism. He asked himself whether nature really needs both these forces. Could it be that in their natural habitat, they automatically become one force, and that we see gravity and electromagnetism as two different forces simply because we are looking at them the wrong way, by *not* getting to see them in their natural habitat in which their laws automatically become more unified and consequently more beautiful? By way of an analogy, think of the cheetah, the swiftest animal on earth, extremely graceful as it runs after its prey in its natural African habitat. Now think of the same cheetah confined to the cage of a Midwestern zoo, pacing incessantly, a truly pathetic creature. Maybe we are looking at the laws of nature in a setting that is not their natural habitat.

What would this natural habitat then be like? There is little we can do to the habitat of the natural laws, short of changing the number of space dimensions. Nordström had the inspired idea to ask what physics would be like if space had four rather than three dimensions. Then, adding time as yet another dimension, as is usually done in physics, we would end up with a *five-dimensional space-time*. Nordström observed that Maxwell's familiar theory of electromagnetism in such a five-dimensional space-time could account for *both* gravity *and* electromagnetism in our customary four-dimensional space-time. To account for our seeming blindness to this extra fifth dimension, he made the *ad hoc* assumption that physical quantities do not change their values as one moves along this extra dimension.

The four-dimensional theory of gravity he deduced by this procedure was identical to one he had discovered already. According to this theory, back in the familiar four-dimensional space-time, gravitational radiation unlike light, cannot be polarized, and light itself is not deflected by the gravitational attraction of the sun.

At the time Nordström introduced these revolutionary ideas, Einstein was in Zurich, hard at work on his own theory of gravitation — the one we call general relativity, a theory in which gravitational radiation *can* be polarized and light *is* deflected by the gravitational attraction of the sun. This theory starts from some deep conceptual problems not even remotely addressed by Nordström in his theory. Accordingly, Einstein, though aware of the theoretical consistency of Nordström's ideas, regarded them as a poor relative of his own ideas. As far as he was concerned, Nordström's proposal was to be kept in mind when testing all these ideas in experiment, but it was hardly worth his attention. Moreover he saw a competitor in the Finn and didn't take well to that. By contrast Nordström admired Einstein and made the long journey from his homeland to the shore of the Limmat to discuss matters with the master of relativity. In what may have been an intentional slight, Einstein refused to receive his Finnish visitor, who left understandably dejected and probably quite angry.

In 1919, Sir Arthur Eddington found that light is deflected by the sun, exactly as predicted by Einstein, and this ruled out Nordström's theory. Nordström graciously accepted defeat and went on to find along with H. Reissner one of the most important solutions of Einstein's equations.

But the story did not end there. After the experimental confirmation of Einstein's general relativity, Nordström's bold idea of a fifth dimension resurfaced in the 1919 work of Theodor Kaluza, who ironically, rather than having gravity of the Nordström type in a four-dimensional space-time correspond to an electromagnetism in a larger five-dimensional space-time, proposed that electromagnetism in our four-dimensional space-time corresponds to a five-dimensional gravity of the Einstein type. After sitting on the fence for two years, Einstein forcefully endorsed Kaluza's theory. Kaluza did not refer to Nordström's earlier proposal. It is both understandable and excusable that, being a mathematician, he was not familiar with it. That Einstein did not correct him is very hard to understand. In all his papers on Kaluza theory, Einstein never once refered or even alluded to Nordsröm's work, the very work that Finn had wanted to talk to him about in Zurich.

Surprisingly, this daisy chain of credit deprival kept growing till it came to an abrupt and sad end. Shortly before World War II, Pascual Jordan, one of the fathers of modern quantum theory, found a hidden assumption in Kaluza's work. He showed that by removing this assumption, one obtains from the unified theory a gravity theory for the usual world with only three space dimensions, which is *neither* Einstein's *nor* Nordström's, but rather a *combination* of the two. The Nordström component of this Jordan gravity could be made sufficiently small to agree with Eddington's results to within the stated errors. Remarkably Jordan did not quote Nordström, of whose work he was apparently unaware.

In 1961 at Princeton University, Bob Dicke and his student C. Brans rediscovered this hybrid theory of Jordan. They too were unaware not only of Nordström's old work, but even of Jordan's much more recent work. Dicke was an experimental physicist of greatest distinction. In his time he had made some famous high precision experiments and has held patents for some spectacular inventions, especially in the realm of radar. Dicke realized that if the sun were more oblate (non-spherical) than theories of the sun would lead us to believe, then the hybrid Brans-Dicke, or more fairly, Jordan-Brans-Dicke theory of gravity, with a predominant Einstein component and a small but *non-vanishing* Nordström component as well, still had a chance. So he set about measuring the sun's oblateness and found it to be large enough to rule out pure general relativity, thus establishing the validity of the hybrid theory. *The New York Times* reported in glowing terms about this discovery and here is where the daisy chain finally broke. When the results of an experiment with such far-reaching implications are announced, the adrenaline of experimental physicists all over the world starts flowing and they repeat it with improved accuracy to make sure its claims are justified. For the solar oblateness experiment this was not the case, the better experiments brought Einstein's general relativity back with a vengeance.

There are many lessons to be learned from the story of Nordström's theory. First of all, we encountered three new examples of the Arnold principle. Second, for a physics discovery a *New York Times* endorsement is often the kiss of death. When it comes to the press it is best to take reporters with a grain of salt — and for pundits get out the saltshaker. Third, let *others* do the experiments which decide the validity of *your own* theory! Fourth, and most importantly: in the end history always gets it right. It may be too late for those concerned, like it was for poor Nordström, who died young. On him lasting pain may have been inflicted, unless somehow he was strong enough to hold on to the belief that sooner or later his time would come.

As for Einstein, his interest in unified theory only increased with time, and unfortunately not for the best, as we saw in the first chapter of this book.

Now that Nordström, Kaluza, Jordan and Dicke are all dead and the daisy chain has long been broken, one might wonder about the fate of the fifth dimension. Remarkably enough, it is doing splendidly. First of all, the

Swede Oskar Klein gave us a better reason to account for the fact that we cannot see this fifth dimension with the naked eye. He suggested that rather than extend to infinity in both directions like the other three space dimensions, this additional dimension curls up into a tiny circle. Then there is no problem with our observations. Imagine beings that live in one dimension, like needles on a line, instead of living in three dimensions, as we believe we do. Traffic problems aside, suppose that these one-dimensional beings develop powerful microscopes, and one day make the epochal discovery that they do not live on a line as they had always believed, but rather on a cylinder, like the skin of a garden hose — except that the circular cross-section of this hose is invisible to the naked eye, being much smaller than an atom or whatever these beings are made of. The circle can only be seen with the help of microscopes. For their everyday existence these beings still have to contend with the nuisance of living on a line. But the laws that govern their existence are constructed to fit the hose on which they really reside.

Now this circle, the hose's cross-section, adds only one such extra dimension, but these beings could be piling up ever more of these small, curled-up "compact" dimensions until at the microscopic level theirs becomes a very richly endowed geometry. By the same token, we, with our three "large" dimensions, could also find at very small scales that our dimensionality keeps increasing. Is there any way of ever noticing these extra dimensions? Yes from our four-dimensional point of view they imply the existence of certain families of particles, known as Kaluza-Klein towers, whose masses are related to the size of the small dimensions. The smaller these circles — or spheres, doughnuts, Calabi-Yau manifolds, whatever — are, the predictably heavier the particles in these Kaluza-Klein towers become.

Then, could we be living in 10,379 dimensions after all and just not as yet be aware of it? *Not really,* if string theory has any relevance in reality. Ordinary particles can move in a space, no matter what its dimensionality. Strings are much more capricious objects and to make sense have to grow their own ten-dimensional space-time. Even weirder is the fact that as the interaction between strings — the interaction that makes one string break into two, or two strings merge into one — becomes stronger, this ten-dimensional space-time, as if on order, grows itself another space-dimension and we reach the maximal total of eleven dimensions. String theory then requires extra space dimensions, but not thousands of them, just a total of ten, so that together with time they yield an eleven-dimensional space-time.

With all these small hidden space dimensions, we may wonder whether there could also be hidden time dimensions. The answer is an emphatic no. Adding an extra time dimension, aside from the upheaval it would cause in the Swiss watch industry, would allow time travel, and land us right in one of the worst neighborhoods of Hollywood.

Finally, I must mention the problem of comparing with experiment the existence of extra dimensions required by string theory. The so very successful *standard model* of modern particle physics — solidly grounded in three space- and one time-dimension — contains numerous arbitrary parameters, like say the ratio of the masses of the electron and its fatter cousin, the muon. These parameters should now be calculable. Trouble is, they depend on the shape in which the extra dimensions curl up. The original expectation was that the equations of string theory will lead to a small number, of such shapes or "string vacua" around which the string theory is to be considered. Intense work should then quickly find what each of these shapes predicts for the ratio of those masses and all the other arbitrary parameters allowing us to nail down the "right shape" of the extra dimensions. Once known, this shape should be explored in great detail and it would surely be found to possess some remarkable features which account for its selection by nature as the only right shape.

This sequence of expectations couldn't be further off the mark. In reality we encounter not a few, but an immense number, a number with something like 500 digits of such string vacua, and not one but millions of these vacua will give acceptable values for some or maybe even all of the standard model parameters. At this point we can throw up our hands and say that around all these vacua possible worlds can arise. We find a whole *landscape* of possible worlds, many incapable of supporting intelligent life and therefore inexplorable. Four centuries later we are back to a much magnified version of poor Giordano Bruno's reasonable if fatal heresy. Not surprisingly, high ecclesiastic personalities are nowadays publicly condemning string theory along with evolutionary biology, stem cell research, and evolutionary cosmology like in the good old sixteen hundreds.

It becomes imperative to find vacuum selection criteria and a lot of work is currently channeled in this direction. It takes a lot of doing even to find criteria that keep the number of usable string vacua finite. Even more refined criteria are needed to get rid of a large region of the *swampland* of vacua where nothing phenomenologically viable results.

But difficulties are both expected and challenging. We only need time and perseverance. Looking back however, in spite of the current dearth of experimentally testable predictions, we have made progress. Not even talking of the immense mathematical and conceptual progress brought by string theory, a consistent quantum approach to gravity and therefore also to cosmology has been developed. Maybe the first test of string theory will come in the realm of cosmology.

27

A Brief History of Space

Galileo and Newton thought in terms of one "God-given" time, experienced by all observers alike, no matter how they moved towards or away from each other. Space was also a given for them, described by the three-dimensional Euclidean geometry of the ancient Greeks. Everything was moving in this space; this just had to be taken on faith, or as an axiom, as the mathematicians call a thing to be taken on faith. This didn't seem like asking too much. After all, that the space in which we move is three-dimensional is intuitively so obvious that at first glance this fact doesn't even warrant discussion. Yet, as we just saw, the future may hold additional dimensions in store for us.

On the other hand, that the space of physics should obey all the axioms Euclid imposed on it, also sounded eminently reasonable, for what else could it have been? Galileo and Newton certainly knew of no alternatives. But then in the nineteenth century the first alternatives were discovered and henceforth it became much less compelling to just take Euclid on faith.

This discovery of non-Euclidean geometries is one of the greatest ever in the venerable history of mathematics. It has a checkered history of its own. In 1823 the 21-year-old János Bolyai, then stationed in Temesvár in the Banat (now Timişoara in Romania, my own hometown) as an officer in the Corps of Engineers of the Austrian Army, discovered that Euclid's "axiom of parallels" is independent of the other axioms of his geometry, and that although in Euclidean geometry through each point you can draw one and only one parallel to a given line, it was possible to construct geometries in which through each point you can draw more than one. Incidentally, there are other geometries in which you can draw *no parallels* at all to a given line — all lines intersect. If this sounds impossible, think of the surface of a sphere, on which the great circles, the ones having the same radius as the sphere itself are decreed to be the "straight lines." It is readily checked that this "decree" respects all the axioms of Euclid except the axiom of parallels. It is now intuitively obvious that there are no parallel lines on this sphere, since any two *great* circles intersect. It is important that we decreed the straight lines to be great circles, otherwise we would encounter smaller "latitude" circles and these can be parallel. The nontrivial fact is that this is not an arbitrary decree, being natural enough to meet all the other axioms of Euclid.

János Bolyai's father Farkas Bolyai, himself a geometer, was preparing at the time the *Tentamen,* a treatise on mathematics. The book appeared in 1831 and included his son's discovery as an Appendix. Bolyai the Elder wrote to Carl Friedrich Gauss, his former Göttingen classmate and friend, then reigning prince of the mathematicians, asking him for his opinion. To claim more than two millennia later that something in Euclid could be changed was nothing short of revolutionary and the Bolyais, father and son, were fully aware of this.

In one of the most despicable actions ever taken by a scientist, the great Gauss replied that he could not praise his friend's son's work for

> *"to praise it would amount to praising myself. For the entire content of the work ... coincides almost exactly with my own meditations which have occupied my mind for the past thirty or thirty-five years."*

Gauss was essentially claiming priority to this discovery. His excuse for not publishing was along the lines that he did not find his contemporaries mature enough to digest his revolutionary insight.

Nikolai Ivanovich Lobachevsky's 1826 work, in which he discovered the same geometry as János Bolyai, was first rejected by the St. Petersburg Academy, but then finally published in 1829 in the *Kazan Messenger*. In due course this paper made it to Gauss, who didn't hesitate to gloatingly inform his friend Farkas Bolyai, in effect that his son had just been demoted from silver to bronze, the silver now going to the Russian and the gold remaining in Göttingen. Bolyai the Younger was furious; he supposedly was not even convinced of Lobachevsky's existence, but was inclined to believe in some foul play on Gauss' part. Lobachevsky was apparently not the world's most careful man, and Bolyai found a host of minor flaws in his paper, which he didn't hesitate to point out. Beyond that however, Bolyai could not cope with the disaster which had befallen him, and his career, along with his physical and mental health all came to a sudden end.

One could say in Gauss's defense that what he did was nothing more than claim the credit he was certain was his due. What's wrong with that? For one thing, even if Gauss did discover non-Euclidean geometry, he did not have the courage to publish this revolutionary finding. In science taking even the most sublime discovery to the

grave, is something that is never, and *should* never be honored — one must have the courage of one's convictions. But
in Gauss's case the story has a darker side as well. This is
not the only such priority claim made by Gauss. Gauss's
other arch-victim was the French mathematician Adrien-
Marie Legendre. In 1806 Legendre invented the method of
least squares for fitting a curve to some available data.
Three years later, Gauss published the same method and
while acknowledging that he had seen it in Legendre's
book, he claimed that he knew the method all along and
deserved credit for it. To this day, most people associate
Gauss's name with this method.

Legendre was the first to ask the question as to the
density of prime numbers among the natural numbers. He
conjectured a one-parameter formula for this density (for
the asymptotic form of this density, to be precise). From
tables of prime numbers, he determined the value of the
parameter "experimentally" as being very close to one.
Years later Gauss took up the problem and made the
improved conjecture that this parameter was exactly equal
to 1. He again claimed that he had known all this ahead of
Legendre, but just hadn't published it. By now, both these
conjectures have been proved.

The fundamental quadratic reciprocity in number
theory, formulated to this day in terms of "Legendre symbols," had received many proofs from Gauss, who then
claimed credit for the idea as well. The bewildered Frenchman
openly vented his outrage when he wrote about Gauss

> *"This excessive impudence is unbelievable in a man who has
> sufficient personal merit not to have need of appropriating the
> discoveries of others."*

Even if one denies Gauss' credit for all the Legendre
and Bolyai related discoveries, he still towers as one of the
greatest mathematicians ever. Yet somehow this supreme

genius could not concede major achievements to anyone; he did not shy away from using his status as a prince of mathematicians to place *droit du seigneur*-like claims where *noblesse oblige* would have been more appropriate. Gauss could not accept that he had, simply, been scooped. In all fairness the word scooped did not exist in nineteenth century German. So maybe Gauss was just "linguistically challenged." In any case his joint statue with Wilhelm Weber still stands today in Göttingen, and celebrates their joint invention of the .... telegraph. To his credit, Gauss never claimed to have invented the Morse code.

This story is a cautionary tale even in today's fast world, and it reminds us again of the wisdom of Plutarch's words "It does not follow of necessity, that if the work delights you with its grace, the one who wrought it, is worthy of your esteem," which I could not resist quoting yet a second time here.

Such rivalries, claims and counterclaims have always been part and parcel of a competitive endeavor such as physics always was and always will be. Think only of Newton's longstanding feud with Leibniz over priority in the discovery of calculus. Today we know that they worked independently of each other and that crucial contributions to this discovery were also made by Archimedes and by Fermat, whose work was familiar to Newton. To this day such incidents occur with amazing regularity. The beauty is that history sorts things out in most cases, so that something closely approximating justice is ultimately rendered, not a careful legalistic form of justice, but rather the curiosity of later generations, unbiased by passion, ambition, and the kind of self-delusion that often plays a crucial role in such feuds.

The very evolution of physics as a science relies on this curiosity of the unbiased young. As in any worthwhile human enterprise, the road to breakthrough in physics is paved with near misses, false starts, and outright mistakes. When the dust settles, it is not that those whose ideas proved untenable suddenly see the error of their ways and accept the arguments of their opponents. Nothing could be further from the truth. As Max Planck remarked long ago, in spite of its eminent rationality, physics does not evolve by having those who are right convince through careful reasoning those who are wrong, far from it. Physics evolves through the many ideas that are out there at a given moment. The physicists who backed the losing ideas hold on to their fallacies, but eventually they die — nature has it all figured out — and the new generation, not having any emotional capital at stake, automatically picks the correct ideas and runs with them.

Back to my brief history of space (pun intended). Euclidean space is flat — think of a Euclidean plane, as flat as can be. By contrast, its non-Euclidean cousins are curved — think only of the sphere, which we found to be such a space and which is clearly curved. But a sphere is uniformly curved; it is as curved at the poles as it is at the equator or anywhere in between. In his general relativity, Einstein proposed that physical space is curved, but not uniformly so: it is more curved where there is a lot of mass, energy, and momentum and less curved where mass, energy and momentum are less plentiful. Einsteinian gravity itself serves as a measure of this curvature.

String theory adds a new twist to all this, for it comes endowed with a smallest length below which it is meaningless to go. Below this smallest length, geometry itself ceases to make sense. This smallest length could be as small as a hundredth of a billionth of a billionth of an atomic nucleus's already small size. Once such a smallest length appears in nature, other remarkable things occur. This had been suggested already in the 1960's by Heisenberg's close friend Carl Friedrich von Weizsäcker.

At the war's end, Weizsäcker, who during the war had had himself appointed to a professorship in Strasbourg, then under Nazi occupation, switched his main focus from physics to philosophy. He founded an institute for the study of the future (*Zukunftsforschung*). This prompted the distinguished experimental physicist Wolfgang Paul's (not Pauli!) quip that only someone who during the war would set his future on the occupation of France would think of himself as amply qualified to study the future. Yet, this move to the philosophical made von Weizsäcker immensely popular in his native land. In a poll he placed right up there with the chancellor, Konrad Adenauer, as one of the two most popular personalities in West Germany. Later on, his brother Richard was to become one of Germany's best presidents. For better or worse, the von Weizsäcker family had been prominent for quite some time. Carl Friedrich's father had been Hitler's ambassador to the Holy See.

Weizsäcker's prewar physics contributions are remarkable for their simplicity and for the clever intuition that drives them. In the Sixties, melding physics with philosophy, von Weizsäcker attempted to follow up on the idea of a fundamental smallest length, then seriously considered by Heisenberg. In the absence of such a length, to describe the "quantum state" of even as simple a system as one particle, one needed to know the probability of finding that particle in no matter how small a region anywhere in space. Very much information is encoded in these probabilities — in fact it is rather obvious that there are infinitely many different possible states, just change the probability ever so slightly here or there. All these infinitely many possible states span what the mathematicians call a Hilbert space, an infinite-dimensional space. Weizsäcker realized that if a smallest length existed, then in a cube in space whose side has precisely this smallest possible length, one can ask only one "yes/no" question: is something there — yes or no? So Hilbert space will no longer be infinite-dimensional. Rather its dimension will be given by the number of all possible sets of answers to all possible yes/no questions. Suppose there were only one such possible question, then this space would have two dimensions. Two questions would mean $2 \times 2 = 4$ dimensions, three questions $2 \times 2 \times 2 = 8$ dimensions, and so on. But how many such questions are there? Obviously, reasoned von Weizsäcker, there are as many such questions as there are smallest cubes in all the universe. Knowing the size of the universe from the astronomers and making the — now rather silly sounding, but in those days seriously entertained — assumption that the smallest length is of the order of the size of an atomic nucleus, von Weizsäcker concluded that the number of smallest cubes in the universe is a number with 120 digits. Multiply the number 2 with itself this many times, he said and you find what nature deals us in lieu of infinity when it comes to the dimension of Hilbert space.

I have never been able to locate the paper in which von Weizsäcker made this claim, but there was no need for that, for he went on talk shows and made widely reported news with it. When confronted with a number as colossal as two to the power 10 to the power 120, a mortal's closest idea of infinity, most of us are awed. In view of the prominence of the personality making this claim, journalists were awed even more. Believe it or not, journalists started flocking to their local physics departments for reaction to what they believed to be a major breakthrough.

In Hamburg, a local journalist approached the leading theoretician, the brilliant Harry Lehmann, a man with an earthy and marvelous sense of humor. When asked to comment, Lehmann replied that he found the idea cute, but would like to suggest a minor correction. He said that in his view the dimension of Hilbert space was only half of the colossal number suggested by von Weizsäcker.

The gullible newsman, hoping for a scoop, asked Lehmann on what he based his assertion.

"According to Professor von Weizsäcker the dimension of Hilbert space gets doubled for every possible yes/no question" Lehmann began, "and among all possible yes/no questions is the one as to whether Professor von Weizsäcker's theory is sensible. But to that question we already know the answer and therefore it would be unfair to double the dimension of Hilbert space on its account." Unfortunately, the journalist did not share Lehmann's keen sense of humor and Lehmann was derided in the press as an envious mediocrity jealous of true genius. If anyone had reason for jealousy here, it certainly wasn't Harry Lehmann. He told me about this on a longish tram ride in Hamburg, and though he was visibly angry about the treatment he got in the press, the twinkle in his eyes was telling a different story. Wild though this von Weizsäcker idea may sound, it has resurfaced over the last years in a slightly altered guise.

The Coming Age

Now that we are approaching the end of the Witten era, it is worth taking a moment's time to assess just how far we have come, where we are headed, and whether there is a new leader in sight. I think the answer to this last question is yes and I am willing to speculate about who this leader might be. I am fully aware of the dangers involved in such speculation — look at poor St. John the Baptist. But then there are no Salomes around in Chicago these days, and by warning you that I am speculating, I am not sticking my neck out all *that* far after all.

String theory evolves through repeated revolutions. The latest string revolution started in the late Nineties with the work of the Argentinean physicist Juan Maldacena. He showed that, in string theory, a world that presents itself to us with gravity, so that in it apples can fall, can also be viewed as a world in which there is no gravity — apples do not fall — and which has fewer space dimensions. It is as if not only does the dimension of space not necessarily have to equal three, but it can change depending on the way we look at the world. He and others gave a full dictionary for translating the physics of one world to the other. Maldacena's idea is reminiscent of holography in which three-dimensional information is encoded in a two-dimensional picture.

This holographic approach is maybe *the* dominant issue now under study, so not surprisingly, Juan Maldacena appears to be Witten's emerging successor. There is also a marked shift towards the interface of cosmology, astrophysics, and string theory, which started when in 1996 at Harvard Andy Strominger and Cumrun Vafa first showed that the very puzzling entropy — calculated long ago by Jacob Beckenstein and Stephen Hawking — of those mysterious objects, the black holes, can be beautifully understood in string theory.

The enterprise of theoretical physics is in good hands, and it is going on driven by the unquenchable thirst for knowledge of the young. It is a worldwide enterprise, and everybody can contribute. It doesn't matter who you are or where you hail from, but you must have something new to say. It is a splendid game and everyone is entitled to play, but you must be serious, you cannot just come and claim preferential treatment.

Here you may rightly expect a story about Maldacena, and I am sorry to admit that I know no such story. He is so much younger, and you only get to hear the stories about your elders and to live the stories involving your peers and your own self. All these stories about the new generation are reserved for the future, when some other physicist will take up where I left off.

Epilogue: The Goal and the Means

Unlike mathematics, physics is a science with a clearly defined goal. In mathematics you have input from physics, from economics, from logic, from biology, from games, and what not — if it catches the imagination of the mathematicians, it is put through their exquisite wringer. There is however no ultimate goal that once attained, exhausts the full purpose of mathematics. I cannot foresee a time when mathematics is done as a science and no new problems remain, just mopping up here and there. I cannot imagine a time when mathematics would become an exquisite corpse, to borrow a term from the surrealists.

By contrast, physics, or more precisely the part of physics that concerns itself with finding the laws of nature, has a clearly stated goal. A few problems will then remain, a theorem here, a corollary there — but as Lagrange complained in the wake of Newton's epochal discoveries, unfortunately the laws of nature are up for discovery only once. It is possible that we'll never be done with this job, maybe the deeper we penetrate to the core of things, the more new laws and surprises await us. Maybe there is no way of even knowing *when* we are done with the job. But in his heart of hearts every theoretical physicist fervently hopes that there is a well-defined end to it all and that he or she will be present at and maybe even assist in the ceremony of bringing these laws down from the mountain.

But what happens after this feat has been accomplished? One view is that we move on to other tasks. With these simple laws all known, we may turn to the study of the very complex on the one hand, and the very useful on the other hand. This would be the era in which cosmology, condensed-matter physics, chaos theory and other related subjects would not only flourish but truly dominate the field. Are we close to this cosmological condensed-chaotic utopia? I wish I knew. Somehow I doubt I'll get to see the Promised Land and that even my grandchildren will get a glimpse of it.

A look at the past evolution of our field gives a better perspective. The first major breakthrough was made in the seventeenth century, primarily by Galileo, Newton, Kepler and Huyghens. Its products are mechanics, optics and a usable theory of gravitation. How many scientists were at work worldwide at the time of this major breakthrough? Fewer than you'd find in or near Boston today.

Next, in the nineteenth century, in the wake of the development of the steam engine in England and France, we had a second breakthrough, a theory of heat, the thermodynamics associated with the names of Helmholtz, Sadi Carnot, Clausius, and a few others. A larger, but by today's standards still modest number of physicists at work worldwide.

The same can be said about our understanding of electromagnetism initiated by the pioneers Coulomb and Ampère and culminating in the nineteenth century breakthrough of Michael Faraday and James Clerk Maxwell.

In the twentieth century, the pace picked up and the number of physicists grew dramatically. The century opened with everybody but a few older people taking atoms seriously — Boltzmann convinced them of that. The older people, Ernst Mach and the chemist Wilhelm Ostwald among them, reproached Boltzmann that no one had yet seen atoms and molecules and one cannot build one's view of the world on an assumption concerning the existence of units of matter which can never be found. Were they ever wrong! Today we have pictures of molecules and even of individual atoms. You'd think that after Einstein's and Bohr's work these reactionaries must have seen the error of their ways and thrown in the towel, but nothing could be further from the truth. It all happened in the Planckian manner outlined in Chapter 25. Mach, Ostwald and a few other influential critics died, and Boltzmann's ideas carried the day.

At this point, the historical account turns more technical.

Once you have atoms, you have very small microscopic objects and the laws of classical deterministic physics no longer apply — the new quantum theory had to be discovered by Heisenberg and Schrödinger. One also runs into high velocities, and this leads to special relativity (Einstein). Once one admits that at velocities close to that of light the laws change, one has to revise one's ideas about gravity, and general relativity (Einstein again) comes into being. But one can have velocities close to that of light *and* microscopic objects, and one needs relativistic wave equations, which predict the existence of antimatter (Dirac). This called for quantum field theory and renormalization (Feynman, Stueckelberg, Schwinger, Tomonaga, Kramers), with the deep meaning of this theory to emerge later in the work of Ken Wilson. At this point I should warn you to fasten your seatbelts, for over the next two pages I will keep up this name-dropping. It isn't absolutely essential that you get all these names, oh

God what am I saying?! In seminar talks theoretical physicists often show complicated looking formulas and warn you that the details are not important, even though without them there would be nothing to talk about in the first place. It is in this sense that you should take my warning. What is essential is that you appreciate the fact that many remarkable people are involved. Now back to the recitation.

The successes of quantum field theory in general, and of quantum electrodynamics in particular, highlight the role of gauge invariance (Weyl, Yang, Mills). This brings symmetries into play, and some of these symmetries have to be broken (Nambu, Goldstone, Brout, Englert, Anderson, Higgs). Then all the particles seen in experiment have to be reduced to a few essential ones, and this leads directly to the quarks (Gell-Mann, Zweig). By now, this is getting unwieldy and these ideas all have to be put together like the pieces of a puzzle, and in a first instance this leads to the *electroweak* part of the standard model (Weinberg, Salam, Glashow). But, is such a standard model consistent? Yes! ('t Hooft and Veltman). And how do strong interactions fit into it? Through color (Greenberg, Han, Nambu, Politzer, Gross, and Wilczek, 't Hooft). Now new chapters of the standard model can be written by 't Hooft and Polyakov. Is this a nice final answer at the microscopic level? Definitely not! It is quite messy and apples haven't yet started to fall, gravity has not yet been included. To include gravity we need extended objects — strings (Scherk and Schwarz, Nambu, Susskind, Nielsen, Witten, Alvarez-Gaumé). For these strings not to contain disastrous instabilities we need supersymmetry (Ramond, Neveu, Schwarz, Gol'fand, Likhtman, Wess, Zumino, Gervais, Sakita). There is also a very important supersymmetric counterpart to Einstein's general relativity (Ferrara, Freedman, van Nieuwenhuizen, Deser, Zumino).

It all sounds so inevitable, so compelling, yet the historic record does not in the least reflect this inevitability. Take string theory, which was born out of some successful strong interaction phenomenology (Dolen, Horn, Schmid, Freund, Harari, Rosner) which led Gabriele Veneziano to a bold guess that turned out to be the cornerstone of string theory. That it was a string we were dealing with, was first recognized by Nambu, Susskind and Nielsen. Many people then picked up this idea of strings, but in the wake of the successes of gauge theory, strings were abandoned by all save a few hardy souls, most notably John Schwarz and Joel Scherk, who argued that if strong interactions can best be described in terms of a gauge theory, then this string theory may serve a higher purpose, a theory of everything and thereby also the first viable theory of gravitation.

The way Gabriele Veneziano discovered the string amplitude is also remarkable for what it reveals about how our field advances. Maybe the best way to make this point is through a literary analogy with Ernest Hemingway's *The Old Man and the Sea*. Santiago, the old man, does not decide "Today I'll catch a big fish" — no, he just goes out in his small skiff and follows the fish-eating birds that are all headed in the same direction. This takes him far out, and his hunch is rewarded with the catch of an eighteen-foot marlin with a "handsome beautifully formed tail." He realizes he has caught something that will be hard to haul in, but instead of letting go, he makes the effort, gets attacked by sharks, holds on, and by the time he makes it in total exhaustion back to shore, all that is left of the beautiful fish are the bare bones. But that is enough, people can reconstruct the marvelous catch by its carcass.

That's more or less what Veneziano did. Following some then much discussed ideas, his birds, he followed them out and came back with one tiny bit of his marlin, but a bit big enough to start constructing the whole edifice of string theory. Things often happen this way. You do not make big discoveries by saying "Today I'll make a big discovery." If you did, your chances for returning with a catch would be negligible, and the wear and tear on your self-confidence could be catastrophic.

The Scherk-Schwarz suggestion, elevating string theory to a theory of everything, sounded so ambitious that most people just smiled it off. But already in gauge theory we had encountered and overcome so-called anomalies (Adler, Bell, Jackiw) and when these anomalies are eliminated from string theory, some spectacular new possibilities arose and this gave rise to the First String Revolution of 1984, its Marats and Robespierres being Schwarz, Michael Green, Gross, Harvey, Martinec and Rohm. At the same time, Candelas, Horowitz, Strominger, and Witten showed us the natural way of connecting string theory to the world we live in. Natural, yes, but experimentally successful? Not as yet.

Since its 1984 revolution, string theory has followed an interesting pattern. Once a major new idea is introduced, the world community of about two thousand string theorists jumps at the problem and within the year there is major movement. Then the work continues, though with the passage of time it keeps acquiring the rocaille of late baroque. You start getting the type of paper that might as well be titled "Everything You Ever Wanted (Or Perhaps Did Not Want) To Know About This Or That Doable Detail." String theorists get fed up with this kind of development, and their frustration can be the mother of their invention so people start to move on and before long the new revolution gets rolling. So we had a few more revolutions associated with the names of Shenker, Douglas, Polyakov, Migdal, Banks, Susskind, Fischler, Polchinski, Seiberg, Sen, and others. The large number of names I mention serves three good purposes. First, it conveys the fact that many people are at work now; second, it does a modicum of justice to these brilliant people and third, it helps keep as low as feasible the number of new enemies the author of this book will acquire by the very act of writing it.

I am not yet done with the revolutions, but first, think of the fact that the large number of first-rate people now at work does not seem to accelerate progress. You would think that there is some extra momentum to be gained from the sheer size of our effort, yet it would be foolhardy to say that things are moving faster today than they were in the days of Sir Isaac Newton, when but a few souls were at it. Somehow progress is inert; you cannot push it along. A major new idea needs time to incubate, until suddenly it infects the whole world with its power, its beauty. Case in point: the Yang-Mills theory appeared in 1953, but not until the mid–Seventies was its relevance recognized.

Could fewer people have brought us to the point in physics where we are now? Very likely they could have, and in a more leisurely manner. Nowadays communication is as swift as one can imagine. With the help of Paul Ginsparg's Internet archives, you can reside in Chile, have an idea while you are staring at the Andes over your morning coffee, work it out by the afternoon, type it up on a word processor and send it in by the evening. Next morning over tea, a theoretical physicist in Kyoto, overlooking the stone garden of Rioanji can read your paper after having printed it out an hour earlier and exalt in its beauty until the moment when he realizes why it won't work in general. By next morning, your idea *is history*. But maybe the dreaded moment when the Kyoto physicist finds the problem does not come, and a few weeks later your paper *makes history* instead.

For better or worse, it's an amazing speed of communication. When I got into physics, preprints were expensive and cumbersome to produce; we got maybe three on a good week. Today, on the Internet, two-to-three hundred is the weekly dosage, if you follow two archives as I do. Stultifying! And yet we learn how to cope. First you look at the authors. If so-and-so is among them, you know through years of experience that you need not read further — you won't miss a thing. If some other so-and-so is an author, better watch out, for this may be it. If you don't know any of the authors, look at the title. If it's one of the "Everything You Ever Wanted ...." papers, forget about it. If it sounds intriguing and rings a bell, then read the abstract. It could be one of those to which your reaction is "I tried that ten years ago and it didn't work, and now he publishes it, wouldn't you know?" cases. Then you can scroll down.

But then, it could just happen every now and then, that it could be one of the "My God, this is the real thing!" goodies. Then you read it and run to your colleague's office for a let's-shoot-it-down session. If by lunchtime the claim in the paper is still alive, then this is very likely to be *it* and whether you like it or not you start thinking about this idea. And there is, there always is a point that has been missed. I tend to sense this by suddenly being overcome by a very uncomfortable ill-at-ease feeling. I meander around, I do routine things, I feed myself, I go swimming, and then suddenly it clicks.

For much higher stakes, Poincaré thought of his famous Fuchsian functions just as he stepped into an omnibus; Kekulé supposedly got the idea of the benzene ring in a dream and God only knows how many more ideas Sir Isaac might have had, had Trinity installed a shower in his rooms.

In experimental physics, large numbers of collaborators are needed to carry out complex and expensive experiments. In theoretical physics, a blackboard and chalk are still the primary resources. Blackboards come cheap and they don't accommodate crowds. Despite the Internet and peer review, the process of discovery is still as solitary as it was at Trinity when Sir Isaac resided there.

Sources, Comments and Some More Stories Masquerading as Comments

Though presenting itself as a whole, this book is built out of narrative atoms: more than 150 anecdotes, vignettes and stories. For some of them my sources are revealed in the telling, but for others I gave no indication as to sources. It would not be in the spirit of this free-flowing book to suddenly go scholarly with a flood of footnotes.

The following list gives sources for the remaining stories, except for a few stories that I have heard more than once, but for which I do not remember the sources, or stories which are so well known that they do not need a source.

To tempt the reader not to leave this section unread, I spiced it with quite a few more stories. In particular I wish to draw attention to the story about Heisenberg's last visit to the U.S. (under chapter 2 below), to Shafarevich's take on the Marshall Plan (under chapter 12 below), to the three Oppenheimer stories (under chapter 18 below), the Touschek and Ziegenlaub stories (under chapter 19 below), to the R.H. Fowler story (under chapter 22 below).

Chapter 1: Einstein Once Removed

The Schrödinger biography mentioned in the text: Walter Moore, *Schrödinger Life and Thought*, Cambridge University Press, Cambridge, 1989

Chapter 2: Heisenberg's Turn

Heisenberg unified theory

In 1974, Heisenberg made what turned out to be his last visit to the United States. He lectured on the East Coast, then in Chicago. He gave us two titles to chose from: a lecture on his unified theory, or a lecture on the history of quantum theory. To make the visit run smoothly we wisely chose the historical lecture and not the lecture on the highly controversial unified theory, and it was simply superb. Held in the immense Cobb Hall, it was a standing-room-only event; I arrived fifteen minutes before the posted time and had to stand. Heisenberg himself had to sit on the steps while he was being introduced. He lectured about Dirac's immense role in the creation of quantum theory. He explained the truly revolutionary nature of Dirac's discovery of antimatter. "Before antimatter I believed the Hydrogen atom to be built of its nucleus, the proton, and of an electron. After Dirac's discovery I could equally well have said it was built of a proton, an electron and an electron-positron pair, or two such pairs, or any number of such pairs, or also a proton-antiproton pair, etc. In short the question that was so central to quantum theory, 'what is a Hydrogen atom built of?' had suddenly revealed itself as meaningless!"

From Chicago Heisenberg went on for a last stop at Caltech, where, unlike in Chicago, he was asked to lecture on his unified theory. It was again a big occasion presided over by Richard Feynman. As related to me by

Thom Curtright, then Feynman's student who attended the talk, Feynman set the tone by introducing Heisenberg with the words "Before the war Professor Heisenberg was the world's most respected physicist." All during the talk Feynman kept interrupting Heisenberg to point out as clearly as possible the many flaws and shortcomings of his work. By the end, Heisenberg was visibly shaken up. Some colleagues took Feynman to task about his behavior.

"We are talking about Heisenberg, not about some first year graduate student," Feynman replied. "If you cannot expect to hear something not obviously incorrect from Heisenberg, then from whom *can* you expect it?"

For more on *Heisenberg at Farm Hall,* see Jeremy Bernstein, *Hitler's Uranium Club: The Secret Recordings at Farm Hall,* Copernicus Book, NY, 2001.

Chapter 3: A True Aristocrat

I heard about *Stueckelberg's resignation and his rehiring* from staff at the University of Geneva and from Valentine Telegdi.

Stueckelberg's dog: that the dog was reacting to his master's uneasiness was suggested to me by Valentine Telegdi

Pauli talks Kronig out of spin: This story is widely known. What is less widely known, is that years earlier, the future M.I.T. president, A. K. Compton, the brother of Arthur Holly Compton, had also suggested a spinning electron in connection with understanding magnetism, though this was at that time a less compelling argument.

Lee and Yang offer Pauli a bet: Pauli's letter to Weisskopf

Chapter 4: The Conscience of Physics

Pauli's three-quarter Jewish heritage: Having first heard about this extra quarter Jewish heritage from Valentine Telegdi, I confirmed it in the ETH library's Pauli internet archive www.ethbib.ethz.ch/exhibit/pauli/ausreise_e.html

Pauli excused from high school to teach at University of Vienna: This story comes from the late Peter Drucker, the well-known business consultant via his son-in-law, my University of Chicago colleague Bruce Winstein. Also Viennese, Peter Drucker attended the same high school that Pauli had attended some nine years before him. By then this, possibly apocryphal, Pauli story had become part of the school's lore.

Pauli nodding during lectures: This well-known story I heard among others from Gregor Wentzel and Vicki Weisskopf, who also gave me the prescription for a "good" lecture: "Start with something that everyone knows and understands, people like to hear what they know. Then say something that only the experts understand, lest you be accused of talking trivia. Conclude with something no one, not even you, can understand, just to keep the proper respect for physics"

Pauli's treatment of Bhabha and other students: I heard these stories from Gregor Wentzel.

Pauli at Purdue: My source is Bob Sachs

Pauli divorce: For Pauli's rival's Bondian name I am indebted to Laurie Brown. For obvious reasons Pauli was not appalled at this to us so evil-sounding name, but he was appalled at his rival's profession "I could see her leaving me not for a bullfighter, but for a .... chemist!" is how he supposedly put it.

Pauli psychotherapy under Jung's supervision: I first heard about this from Gregor Wentzel, according to whom the Fierz family had a lot to do with getting Pauli to see

Jung. Chris Schmid first mentioned to me that the dreams analyzed in Jung's book are Pauli's.

Pauli effect demonstration: See Sir Rudolph Peierls, Biog. Mem. Fell. Roy. Soc. London **5**, 175 (1959).

Chapter 5: The Wizard of Pasadena

Feynman polaron paper. Years after Richard Feynman told me this story, I got to meet Herbert Fröhlich in Kyoto at the conference on the occasion of the fiftieth anniversary of Yukawa'a meson. He gave a talk on his work with Nick Kemmer on early meson theory. The main point of the talk was that they had told Homi Bhabha (the one mistreated by Pauli) of their work and that Bhabha then also wrote up what they had told him and undeservedly shared in the credit. People don't forget easily about such things, and it is interesting how even in advanced old age they go on passionately holding the grudge.

Feynman lectures on work of his student: this story reminds me of Gregor Wentzel's oft-repeated remark "It is hard to sponsor doctoral students. First *you* have to write them a thesis and then you have to convince them that *they* wrote it."

Chapter 6: Back to Maupertuis

J.M. Jauch told me about *Galileo's canonization manquée*

Chapter 7: The Language of God

At the 1979 Einstein Centenary Conference in Jerusalem I heard a lecture about *Einstein's model of physics research*, which appears in a letter written by Einstein. Unfortunately this lecture does not appear in the Proceedings of that Conference. Such things happen...

Chapter 8: Emmy Noether and the Urge for the Abstract

Concerning the *details about Emmy Noether' life*, see Auguste Dick's interesting biography *Emmy Noether,* Birkhäuser, Basel, 1981.

The *"Der, die, das" story* is widely known. Though the *"das"* part figures in many an oral telling, it does not appear in Dick's biography and may be apocryphal. I first heard it from the mathematician Ion Bîtea at the West University of Timişoara in Romania some time in the Fifties.

Teichmüller demands end to Landau course: I first heard this story from the late Saunders MacLane, who was studying in Göttingen at the time. When Hitler came to power Saunders' first action was to hide his copy of *Das Kapital* under his bed, with the assumption that Hitler's goons would never look there. Then his teacher, Paul Bernays, a Swiss Jew was fired, and Hermann Weyl became Saunders' somewhat reluctant thesis sponsor. He did not see eye to eye with Hilbert and Bernays on foundations of mathematics problems.

Noether at Bryn Mawr: I should add that she got this job with the help of the Institute of International Education. The man who processed her case was the famous journalist Edward R. Murrow, who started his career working for the IIE. I owe this detail to a lecture of Irving Kaplansky.

Hermann Weyl's obituary of Emmy Noether appears in Scripta Mathematica III, 3 (1935) 201 and is reprinted in Dick's book. So is *Albert Einstein's Noether obituary*, which appeared as a letter to the editor in *The New York Times*, May 5, 1935.

Chapters 9: Oswald Teichmüller and Nazi Science

Teichmüller and Bieberbach in Berlin: W. Abikoff, Mathematical Intelligencer **8**(3), 8 (1986) contains a marvelous and well-researched account of Teichmüller's life.

Bers quotation of Plutarch's "Pericles": See the fascinating MacTutor History of mathematics archive biographies of mathematicians (henceforth referred to simply as MacTutor). These have been very useful to me for checking known facts, or as in this case for learning new ones.

Hasse-Cartan exchange [I can't locate my source]

John Thompson comment on Ernst Witt: private communication from John Thompson who spent a summer in the Seventies visiting Chicago where he had gotten his PhD and lectured on "monstrous moonshine," a gorgeous piece of mathematics later destined to show up in string theory.

Jordan-Pauli exchange following publication of Jordan's book: I heard this story from Peter Bergmann who told it to me after a colloquium I gave at Syracuse University in the Eighties, in which I quoted Jordan's theory of gravity, the one later rediscovered at Princeton by Brans and Dicke. I had called the theory the Jordan-Brans-Dicke Theory, nomenclature that was still new then.

Chapter 10: Spontaneous Breakdown of Human Decency

The novel in question is Bernhard Schlink's *The Reader,* Viking, New York, 1998.

Chapter 11: Scientists in Politics

I heard the story about Einstein and the dueling student from Dr. Laci Steiner, a family friend, who had attended the event three decades earlier. A freshman at the time, I was just reading Einstein's gem of a book *The Meaning of Relativity* and was completely under its spell. This Berlin story hit me as if I had been reading Homer and someone had told me "Oh, Homer, that old feller, I met him once in Athens on the Acropolis and he told me...."

Chapter 12: Jews in Science, the Backlash

George Birkhoff's anti-Semitism: MacTutor, though Saunders MacLane has a different take on it.

Yale recommendation letter for Murray Gell-Mann: I heard this story from someone who in turn heard it from Arthur Wightman.

Rowland as expert witness: I heard this story from Ed Purcell during a marvelous dinner at the Telegdis.

Pontryagin autobiography censored: I saw the mutilated issue of Uspekhi

Pontryagin's first wife was Jewish: that she was intensely disliked by Pontryagin is clear from his autobiography; that she was Jewish, I have heard repeatedly, though I know neither her name nor her life story.

Kolmogorov-Aleksandrov relation: I first heard about this from Felix Browder, but, of necessity indirect, comments to this effect are contained in the Kolmogorov centennial volumes: *Kolmogorov in Perspective,* American Mathematical Society, 2000, History of mathematics v. 20; *Kniga o Kolmogorove,* Fazis/Miros, Moscow, 2000.

Igor Shafarevich's "Russophobia" unfortunately enjoys a much wider circulation than any of his epochal mathematical papers. A funny, and at the same time sad, small detail is Shafarevich's assertion that any American whose name is Marshall is sure to be Jewish. He then states as fact that Gen. George C. Marshall was a Jew, and that the immensely successful Marshall Plan to which his name is attached, was part of an anti-Russian conspiracy hatched by international Jewry. Had Shafarevich but paid attention at the very least to the difference between the use of Marshall as a first and as a last name... It's non-commutative, you know.

Chapter 13: Stalin and the Quantum

Fock makes the Copenhagen interpretation of Quantum Theory acceptable in the USSR: In the early Nineties, after the collapse of the Soviet Union, I visited and lectured at the Institute for Theoretical and Experimental Physics in Moscow. There I met Fock's grandson, a young theoretical physicist, and he confirmed to me Loren Graham's version of this story.

Chapter 14: Mr. Sakharov Goes to Kiev

Landau quip about Zel'dovich: I heard this story from V.L. Ginzburg, during an evening get-together at Chandra's then lakeshore apartment.

Chapter 15: Mr. Yang also Goes to Kiev

Jeremy Bernstein's profile of Lee and Yang: appeared in *The New Yorker*, 1962.

553 plus or minus 1 authors of a paper: Eli Rosenberg made this amusing remark during a talk at an Enrico Fermi Institute seminar about an experiment of the collaboration identified by the acronym DELPHI at the CERN laboratory in Geneva.

End of the spectacular Lee-Yang collaboration: the version in the text I heard from André Weil. Weil's own life story is as dramatic as any and would fit marvelously in this book, but it has already been told with unsurpassed mastery by Weil himself in his book *The Apprenticeship of a Mathematician,* Birkhäuser, Basel, 1992.

André Weil is the younger brother of the philosopher Simone Weil. He told me that their parents had taken the advice of his business-savvy uncle and before World War I put all their savings in Russian Imperial bonds, which became worthless at war's end. When he and his sister inherited them, they used them to wallpaper a Paris apartment; they were very colorful after all. After the fall of communism the Russians decided to honor these bonds at some fraction of their nominal value. I doubt, though, that they could be peeled off the wall.

Chapter 16: The Stone Age and the Fermi Era

Fermi as the preeminent twentieth–century physicist to have made breakthroughs in theoretical, experimental and applied physics: I have used the word "preeminent" rather than "only" since Heisenberg's Nobel Prize laureate pupil Felix Bloch of Stanford University also qualifies on all these counts, as stressed to me by Valentine Telegdi.

Chandra at Fermi's deathbed: I heard S. Chandrasekhar tell this story more than once in a voice filled with emotion and with an expression of deep admiration on his face.

Chapter 17: The Serene Sensei

Nambu "spies" on Tomonaga: The story comes from Yoichiro Nambu

Ziro Koba's Quantum Electrodynamics calculation mistake: Yoichiro Nambu first told me this famous story. Koba ultimately left Japan altogether and went to Copenhagen, where, not being a Danish citizen, he could not be appointed as a professor at the university in those days. So he was given two jobs, which made for a financially acceptable arrangement. He had a brilliant Danish student Holger Bech Nielsen, one of the founders of string theory. Koba and Nielsen wrote two famous papers together and then Koba was supposed to travel somewhere in the Third World to present a paper. As a precaution he got the requisite vaccinations, which killed him.

Nambu visits Einstein in spite of Oppenheimer's prohibition: I heard this story from Yoichiro Nambu.

Chapter 18: Oppenheimer, Hero or Antihero

Oppenheimer and the superconductivity preprint: This story is based on my own experience, but similar stories about Oppie's superficiality abound. Here are two I got from David Fairlie. He heard the first one from Abdus Salam. Abdus had given Oppie a copy of his famous paper on overlapping divergences and realized that he had forgotten to include the diagrams, without which the paper would be incomprehensible. So he trotted off to Oppenheimer's office with the missing diagrams. Just then Oppenheimer was entertaining a visitor to whom he explained how he had enjoyed reading the paper of this brilliant young Pakistani physicist who had joined the Institute. Salam said, "Here are the diagrams. You need them to understand the paper!' Oppenheimer turned red and said, 'I think it is perfectly possible to understand the paper without the diagrams!'"

The other story along the same lines David Fairlie got from the late Trude Goldhaber. Oppenheimer was giving a talk himself, and was interrupted by an Indian visitor who said "I don't understand where you got that equation from." Oppenheimer grandly rounded on him "Well, if you don't understand *that*, you should not be here!" At this point the great Eugene Wigner, hardly a novice, produced his signature apologetic cough and seconded the Indian "Excuse me Dr. Oppenheimer, I don't understand it either!"

I should add here that I deliberately stayed clear of the oft-told story of the hearings in which Oppie lost his security clearance. There is one less-oft-told story. When things started getting hot for Oppie and his brother Frank, Oppie sacrificed some of his students on the altar of self-preservation by denouncing them to the authorities as potentially dangerous communists. They were communists, true enough, but harmless ones and to voluntarily denounce your own students, your academic children, when you have something much more serious of your own to hide, is a form of human behavior hard to reconcile with any moral code. As a consequence of Oppie's denunciation, the brilliant David Bohm was indicted and then summarily fired from his faculty position at Princeton University. Einstein, who fully appreciated Bohm's brilliance, demanded that he be given an appointment at the Institute for Advanced Study in Princeton, where Einstein himself was the most distinguished faculty member. But Oppenheimer, by then director of the Institute, vetoed Bohm's appointment. Bohm had to flee the country and in Brazil, Israel and England came up with major discoveries, which far outshine his teacher's contributions. I wonder what Bohm would have produced had he peacefully lived out his scientific life in Princeton. And if you confront this with the legend.... A much more detailed account of this episode is contained in F. David Peat's remarkable biography of Bohm *Infinite Potential*, Addison-Wesley, Reading, MA, 1996.

Chapter 19: Viennese Physics, Then and Now

Hans Thirring's proposal for Austria's disarmament: Hans Thirring wrote me a letter about these events

Touschek friendship with Gina Lollobrigida: unfortunately I couldn't confirm this quite widely circulated story, as I haven't had the opportunity to meet Ms. Lollobrigida, whom I admire very much.

Touschek at Pauli's deathbed: I heard this story from Touschek in an outdoor restaurant in Pisa during a detour in a drive up from Rome to the Garda Lake. Our table was tilting so badly that our plates were sliding around. We asked the waiter to fix it, but he shrugged it off with "Gentlemen, you are in Pisa." Touschek also told me another Pauli story. Once Pauli came to Rome and in the evening wanted to go to a particular restaurant where years earlier he had had a very good Chianti. He didn't remember the restaurant's name or address and they were just walking "on a hunch." Touschek used this time to tell Pauli about his work on time reversal, the one for which he later got arrested. Pauli was nodding and approving "That's right, yes that is right." Touschek, familiar with Pauli's critical nature was extremely pleased. Finally they found the restaurant and Pauli turned to his young colleague "Didn't I tell you all along that we were on the right way?"

Bogoliubov sends crackpot to Landau: Reli Ziegenlaub is the source for this story. In the mid Fifties Reli had been a student in Bogoliubov's solid-state group in Moscow and was finishing her thesis, when in the wake of a politically incorrect — by Soviet standards — statement during the 1956 Hungarian Uprising, she was sent packing home to Romania with the darkest blots on her record. I myself was under a very dark cloud on account of my participation in a 1956 anti-Soviet demonstration. With time on our hands, we read through Sir Rudolf Peierls' beautiful *Quantum Theory of*

Solids, whence much of *my* solid-state knowledge comes from, and together with a number of other local physicists (the late Mircea Zăgănescu, Imre Hegedüs, Tibi Tóro and Otto Aczél) through some textbooks on quantum field theory. In the summer of 1959, Reli went on a tourist visa to the Soviet Union, and while there, contrary to my expectations she did not graduate, but instead married a man active in the Soviet film industry. Later that year I managed to "get out" of Communist Romania and Reli was to return to the USSR and try to work as a theoretical physicist there. Unfortunately, I couldn't remember the name of her husband and thus couldn't track her career, were she to have been publishing under her married name. Actually she didn't, as I later found out. In Chicago I would scour the Soviet solid-state publications for any articles published under her maiden name, but to no avail. At that 1970 conference in Kiev attended by Yang and Sakharov, I ran into her thesis sponsor and asked him for news about her. He denied ever having known her. After the collapse of the Soviet Union, I learned that she had divorced, and had been able to leave the USSR for Israel, where in her fifties she tragically died of cancer. She was very talented. What a waste! As a confession on my part, Tatiana, the central character of my short story *Exotic Spheres,* which appeared in *ACM* issue 40, pp. 117-130, owes a lot to Reli.

Chapter 20: The Importance of a Sense of Humor

Stories about Fritz Houtermans, one of the most colorful and dramatic physicists ever, are legion and have been told in many places. I heard some from Walter Thirring, Reinhard and Mafalda Oehme, Laurie and Brigitte Brown, among others. Some of the stories told here appear also in Thomas Powers' interesting book, *Heisenberg's War* Alfred A. Knopf, New York, 1993. I wish to draw attention to Iosef

Khriplovich's marvelous article on Houtermans in the July 1992 issue of *Physics Today.*

Chapter 21: The Russian Style

Landau-Lifshitz book story: To this day this remains the most remarkable treatise on physics. It was published in installments, as it was written. The graduate students at Landau's institute had to take a candidacy exam on all existent volumes of Landau-Lifshitz. Whenever it became known that a new volume was about to appear, all those who had not yet taken the exam rushed to take it before the new volume appeared and was added to the material required at the exam.

Landau's arrest. I heard the famous story about the guard admonishing Landau, from Gábor Domokos.

Many reminiscences about Lev Davidovich Landau have been written over the years. I used those of V.L. Ginzburg, I.M. Khalatnikov and especially A.I. Akhiezer, all of which have appeared in *Physics Today* in the years 1989 and 1994.

Chapter 22: Chandra, Passionate Cambridge Gentleman

Chandra-Eddington conflict: More on this can be found in K.C. Wali's excellent biography of Chandra, K.C. Wali, *Chandra, A Biography of S. Chanrasekhar,* U. of Chicago Press, Chicago, 1991.

R.H. Fowler: As I said in this chapter, R.H. Fowler, Chandra's teacher at Cambridge, had been the first physicist to bring the then young quantum theory to bear on the problem of stellar evolution. In recognition of this work the Royal Society was to bestow on him in 1936 a prestigious medal. Though not accompanied by any monetary reward, this medal turned out to be one of the most valuable prizes ever awarded to a scientist. It was the only medal or coin ever minted with Edward VIII's likeness on it and as such acquired immense numismatic value over the years. After receiving the Nobel Prize in 1983, Chandra laughingly told me that R.H. Fowler's much more modest medal was by then worth more than his own Prize, while adding that Fowler had donated it to the Royal Society.

John Eaton accompanies Einstein on the piano: The composer John Eaton told me this story at a party following a Mandel Hall concert, at which my voice teacher Elsa Charlston sang some of his remarkable songs.

Chapter 23: On and Off the Map: Romanian Mathematics

Stoïlow's death: The well-known Romanian topologist Valentin Poenaru, one of the two fresh holders of the doctor degree hosting the fatal dinner told me this story. The other one was Ciprian Foiaş, as far as I recall.

Ceauşescu abolishes mathematics section of Romanian Academy: I first heard this story from the Romanian physicist E.M. Friedlander shortly after his arrival in the U.S. in the Seventies. Since then I heard it many times. It was common knowledge in the American mathematics community.

Chapter 24: Du côté de chez Telegdi, the Experimental Side

Telegdi-Wu conflict: Along with C.S. Wu, Marie Curie and Lise Meitner come to mind as the remaining members of the triummulierate of leading experimental physicists.

Maybe here I owe a bit more on Madame Curie, the only person ever to hold Nobel Prizes in both physics and chemistry. The prize in physics she shared with her husband Pierre Curie and the later chemistry prize came her way in the wake of a campaign of defamation aimed at her in the French media. After Pierre Curie's death (he was run over by a horse-drawn carriage, but was very depressed at the time, very likely due to radiation sickness, the first documented case of this disease; his depression was blamed on his wife by the media), his 38 year-old widow and his married former pupil Paul Langevin fell in love. Mme Langevin sued for divorce and named Mme Curie as correspondent. The press had a field day, along the lines "Foreign home wrecker destroys good French couple's marriage." One journalist went as far as calling Langevin "*le chopin de la Polonaise*" which translates as "the Polish woman's meat chop," but in French comes out as a wicked pun on Frederic Chopin's Polonaises. Things went so far that in the twentieth century (!) Langevin had to fight a bloodless duel to defend his and Mme Curie's honor. It is at this point that a number of scientists, Einstein and Lord Rutherford amongst them, campaigned for a second Nobel Prize (in chemistry), not only for new achievements, but for putting the French on notice that Marie Curie was a woman of genius who should be left in peace. It worked.

Chapter 25: Leader of the Pack

The views expressed in this section are my own.

Chapter 26: New Dimensions

Nordstöm-Einstein conflict: My source on this story is Chandra from whom I also received a photo of Nordström, which Tom Appelquist, Alan Chodos and I wanted to reproduce in our Addison-Wesley monograph *Modern Kaluza-Klein Theories*. Though it took quite some doing to persuade Chandra to lend us the photograph — according to him the last such photograph in existence — it was lost on its way to the publisher, apparently leaving poor Nordström imageless in the annals of science. Fortunately, in his native Finland a further "last" copy of this passport photo did exist, and is now available even on the web. The version of the Nordström-Einstein relation I heard in Finland from Christofer Cronström and Claus Montonen is rather different from the one given in Chapter 24 above. Yet, upon more careful consideration, the two versions are not incompatible. According to this other version, the two men were friends and had good discussions in Zurich. Paul Eherenfest supposedly also participated in these discussions. In 1917 Einstein, the leading theoretician at the University of Berlin, is said to have supported a Berlin offer of an associate professorship to Nordström, which Nordström eventually turned down. But by 1917 general relativity was complete and Nordström himself had found an important solution of its equations. One can relate to someone who is developing one's ideas in a significantly different manner from the way one relates to a serious competitor. Nordström's daughter Saga told me that Einstein had written letters to her father with the salutation "My Dear Friend." But then, human relations can and often *do* fluctuate.

The string theory landscape is beautifully discussed in Leonard Susskind's book *The Cosmic Landscape: String Theory and the Illusion of Intelligent Design*, Little, Brown, NY, 2005.

Chapter 27: A Brief History of Space

Gauss' behavior towards the young Bolyai and the old Legendre: See also MacTutor.

I have heard *Wolfgang Paul's take on von Weizsäcker's study of the future* from Valentine Telegdi.

Chapter 28: The Coming Age

Juan Maldacena is currently a professor at the Institute for Advanced Study in Princeton.

Index